The Digital Challenge for Libraries

THE DIGITAL CHALLENGE FOR LIBRARIES

Understanding the Culture and Technology of Total Information

Ralph Blanchard, Ph.D.

iUniverse, Inc.
New York Lincoln Shanghai

The Digital Challenge For Libraries
Understanding the Culture and Technology of Total Information

Copyright © 2005 by Ralph R. Blanchard

All rights reserved. No part of this book may be used or reproduced by any means, graphic, electronic, or mechanical, including photocopying, recording, taping or by any information storage retrieval system without the written permission of the publisher except in the case of brief quotations embodied in critical articles and reviews.

iUniverse books may be ordered through booksellers or by contacting:

iUniverse
2021 Pine Lake Road, Suite 100
Lincoln, NE 68512
www.iuniverse.com
1-800-Authors (1-800-288-4677)

ISBN-13: : 978-0-595-35069-8 (pbk)
ISBN-13: 978-0-595-79775-2 (ebk)
ISBN-10: 0-595-35069-0 (pbk)
ISBN-10: 0-595-79775-X (ebk)

Printed in the United States of America

Acknowledgements

I would like to thank my friend Professor Doug Garnar, who read sections of my manuscript pre-publication. Professor Garnar has served as Chair of the History/Philosophy/Sciences Department at a major two-year college for twenty-four years (reelected every three years by his colleagues) and a local School Board Member for nine years (one as Board President.) His vast experience with high school and college students enabled him to ask some penetrating questions that were helpful to me. Professor Garnar has dedicated his life and career to students and what we teach them. More people should do what he has done. I would also like to thank Michael Carey, Project Manager/Lead Engineer for Panacea Technologies, Inc., a software firm that specializes in automation and validation services for pharmaceutical and biological batch manufacturing. He is also an Adjunct Professor at a large university in Philadelphia and has taught undergraduate courses in Information Systems and graduate MBA courses in Database Management and in Human Computer Interface Design. As a pre-publication reader, he made some insightful comments and suggestions regarding technology issues covered in the book. A special thanks to Bunny Hoest who graciously gave me permission to use her *Laughparade©* cartoon and who shared with me part of her personal philosophy of life—every important discussion should start with a giggle. The Jim Borgman cartoon is copyrighted and reprinted with the special permission of King Features Syndicate. I also want to thank John Meador and Laurie Miller of Binghamton University for pointing me toward this project. I am donating all royalties from the sale of this book to the Binghamton University library system. I know John and Laurie will invest the money wisely. Lastly, Beazle Louise Blanchard showed great interest in this project and altered and deleted text and graphics on an almost daily basis. I remained calm and finished anyway. Thank you, Beazle.

I want to say that any problems with content, including errors, omissions and fuzzy thinking are my responsibility and should not reflect poorly on those who encouraged me.

Please feel free to visit the companion blog page to this book where you can join the discussion and make your views known. The future of libraries is worth talking about. The address is http://digital-library.blogspot.com. You can also visit http://www.lisfeeds.com/ a web based RSS headline aggregator that currently scrapes almost 200 library-related blogs.

I'll see you at the library.

Contents

1. Introduction 1
I love libraries 1
College Daze 2
The Digital Challenge Ahead 5
The Future of Libraries 8
    It's Not a Problem, At Least Not Here 9
    There is No Money in the Budget 10
Success Strategies 12

2. The Development of Information Services 15
The Digital Economy 15
The Inner Dynamics of a Service Business 17
A Sense of Urgency 17
How to Lose Business by Laughing at Customers When They Need Your Help 18
A Willingness to Take Risks 18
A Case of Unintended Consequences 20
Understanding What Kind of Services Satisfy Customer Need 22
Standardized Service: No Information Content 22
Differentiating Standardized Service With Information Content 23
    Why Are There So Darn Many Starbucks Coffee Shops and How Do They Get People to Pay That Much for a Cup of Coffee? 25

Information Service: Selectively Variable Content, Random Customer Access 28

Information Providers and Information Consumers 30

3. Internet Search Engines 32

Academic Search the Google Way 33
    Auctioning Answers 33
    Battle of the Giants: Libraries vs. Google 35
    Planning a Rematch: Can Libraries Win? 37

Yahoo as a Research Tool 38
    The Tendency of Keyword Search to Deliver Commercial Information 39
    Not Knowing Where to Start 40

Book Search with Amazon's A9 41
    Update: Google Bytes Back 43

4. Search Engine Problems 44

Offline Content: Inaccessible Analog Information 44

Offline Content: Inaccessible Information in Local Systems 45

Offline Content: Inaccessible Digital Information 46

Online Content: Reliance on Keywords 47

Online Content: Identifying Useful Information 48

Blinkx 49

Snap 50

Clusty 50

Online Content: The Impermanence of Information 51

Online Content: The Accuracy of Information 53

Online Content: The Issue of Searcher Privacy 53
Online Content: Satisfied, Overconfident, Uninformed Searchers 54

5. The Role of Advertising in the Search Engine Business Model 57
Money for Nothing and the Clicks Are Free 58
The Influence of Unidentified Advertising on Search Results 59
The Search for Students 62
Information vs. Knowledge 65
Information About Information 69

6. Ambient Information 71
Wi-Fi 74
    The "Broadband Gap" 75
    The Information Politics of Telcos 76
    Local Area Private Wireless Broadband Networks 79
Music Technology 82
    Its Only Rock and Roll but We Like It 83
    No Such Thing As a Copy 85
    Heard in the Dorm 86
Peer-to-Peer Networks 89
    How does P2P work? 90
    Should This Technology Be Controlled? 91
    Suing John Doe 93
    A Vision of Things to Come 96
    "Legalizing" P2P 97

7. Digital Students 100
Understanding Millennials 102
    A Digital Career 104
    Testing for Digital Literacy 105

8. The Library as a Service Business 108
The Non-Profit Budget Crunch 108
    Profit vs. Value 110
Students as "Customers" 112
    Shopping for Courses at the Academic Mall 113
    Losing Student-Customers 114
Developing Digital Personnel 115
Better Accuracy-Greater Productivity 116
Wal-Mart as a Productivity Model 119
    Strategy 119
    Tactics 120
    Service-Oriented Employees 120
    Single-Touch Productivity 121

9. Training 123
The Multiplier Effect 123
Changing Spaces 124
    Changing Spaces at Binghamton University 125
Peer-to-Peer Training 126
A "Gooey" Interface 127
    Training the Apple Way 127
        1984: Orwell Was Wrong 127

    Learning How to Use a Mac 128
    The Graphical User Interface (GUI) 130
    A Future Without Words? 133
Accelerating Change Through Turnover 135
Employee Turnover in Academic Institutions 136
Service Company Hiring Profile 136
    Attitude 138
    Aptitude 139
    Hiring Profile Summary 142

10. Promoting Service 143
Buzz and Viral Marketing 144
    Make it Free 145
    Make It Memorable 147
    Be Sure It Is Networked 148
Interviewing Tactics: A Picture Worth Thousands of Words 149
Push Marketing 151
The Joy of Email 153
    Above the Horizon 154
    Keep It Small 155
    Did It Get There? 156
    Avoiding Spam Filters 157
RSS as an Email Substitute 159

11. Conclusion 161

Index 181

1. INTRODUCTION

©2005. Reprinted courtesy of Bunny Hoest.

The "business model" of a library has remained unchanged for millennia — temporary access to content at no charge. This model is threatened today.

I love libraries

I love libraries. I always have. I don't mean a specific library. I mean pretty much ANY library. One of my most pleasant childhood memories is of the little community library in my hometown. It was in what had been a private home and probably had all of 5,000 volumes and just a handful of periodicals, but as soon as I learned to read it became an almost magical place for me. There was an alcove that overlooked a lake and I could sit there for hours and read about a larger world. It was staffed by quiet, elderly ladies, probably volunteers, who spent their time dusting book jackets or sitting at a small desk near the front door. The library catalog — Dewey Decimal System — was kept on 3 x 5 cards in a box on the desk, and they posted a carefully-typed list of library members with overdue books, sometimes as many as ten names, right on the door where everybody could see it. I made sure my name was never there.

John Dewey invented his Dewey Decimal System not a mile as the crow flies from our community library. He had a small library of his own, sort of a laboratory for classification, set up at a local private residence club (where he lived during his final years.) It showcased his Dewey Decimal System as well as a renowned collection of books on philosophy. I visited this library numerous times, as did many local residents. We rubbed shoulders with visitors making pilgrimages from around the world. One could pull books from the shelves with the classification numbers written on the spine in Dewey's own handwriting. Dewey was an international phenomenon, an intellectual superstar. He passed away in 1952, too soon for me to actually meet the great man. Recently, a collector bought the library's remains including the rights to the classification system itself. There is a web site that promotes the use of the system under the slogan "Do the Dewey!" You can also order a "What Would Dewey Do?" bumper sticker or t-shirt.

College Daze

As an undergraduate at an Ivy League school, I received a student aid package that included a part-time job. The options were, buss tables at the student center or work at the university library. The choice was obvious; the pay was $1.20 per hour, minimum wage. I worked 20 hours per week. I started out frisking student book bags and seizing volumes that were being removed from the building without the proper documentation. It was unpopular work, but it had some great fringe benefits, including a stack pass for the graduate library. I soon had my own carrel on the 6th floor and free run of one of the best libraries in the East. Because I had regular access to great reference librarians, I learned about many obscure sources and research shortcuts. It showed in my term papers and research projects, which were really good (if I do say so myself.) By my junior year I was taking graduate courses and holding my own, at least research-wise, with students writing Master's Theses. When I got to the graduate school at Binghamton University in upstate New York, my library research skills were honed to perfection.

I would have never been able to complete my doctoral dissertation had it not been for the Binghamton University library. Like most advanced degree candidates, I selected an interesting but obscure topic only to discover that virtually none of the primary documents I needed were available in the United States. Upon hearing this, the library reference staff sprang into action. They first located the necessary materials in the British Museum in London and in other European libraries and then made special arrangements to have them microfilmed and sent to me for my research. It took months and cost thousands of dollars, and while I had the time I definitely lacked the money. Not a problem, I was told; the library would pay for everything! Upon completion of my work, the microfilm became a permanent part of the graduate research collection in Russian Studies, where it remains today. I have always been extremely grateful for the library's assistance and generosity. I couldn't have done it without them.

It might sound a little strange to undergraduates today but until recently a university library was cool, "in," a happening place. Socially, it was the equal of any student union or dorm lounge. During my time as an undergraduate, everybody who was anybody in the liberal arts school hung out at the library from Sunday through Thursday and even on weekends during exams and when big term papers were due. There was even a "cigar room" for future CEO's. Like the private clubs they eventually planned to join, it had rules. The most important one was, absolutely no pot.

The library was also a major center of campus academic life and, with the possible exception of some of the lab sciences, of intellectual ferment as well. There was often a line of students waiting at the undergraduate Reference Desk. At Binghamton, in the era before the new Bartle building, the lobby of the library was a center of on-campus activity. Many professors had their offices in the tower and used the elevators in the lobby. As a result, the ground floor was frequently jammed with graduates, undergraduates, Faculty, all chatting about ideas and books and politics. The whole place had a vibrant, intellectual atmosphere, made possible by lots of hot coffee and very slow elevators. Many of the impromptu faculty lectures and student debates were absolutely great. It is one of my best memories of grad student days at Binghamton.

You never knew what might happen at the library. When I was an undergraduate some students let themselves into the building after hours one Saturday night and somehow managed to hang a Playboy centerfold about 30 feet up on the wall directly behind the Main Information Desk. It was discovered when the library opened the next morning and, since there was no way to get it down without a maintenance crew with ladders or scaffolding, it hung there for a day or so. By mid-afternoon the buzz had spread all over campus and the line to get into the building to see what was going on snaked out the door and onto the arts quad. Campus police had to be called in to maintain order. It wasn't the fact that the babe was nekkid; the picture was too high up to show much detail. It was more the rampant speculation as to who, why, how could anybody pull this off? (No pun intended.) Scores of students came in just to take pictures; the place was a zoo. The perps, of course, were never identified but the library's daily attendance skyrocketed and for some reason stayed that way for the next several months. The head librarian, a really great guy and visionary administrator, always made a point of telling an elaborate version of the story to visitors (except parents) and, come to think of it, seemed pretty smug about the whole thing from the moment it was discovered. It makes you wonder what he knew.

On another occasion, stray dogs, which were often let inside the library on cold winter days by sympathetic students, had a disagreement over choice sleeping turf in one of the main reading rooms. Two St. Bernard's, monster dogs weighing a couple of hundred pounds each, growling menacingly and snapping at each other with bone-crushing mandibles, became so enraged that they lost all control. Chairs, even tables, were tipped over and students fled for their lives. It was like one of those Japanese films, Mothra vs. Godzilla, with the girls screaming and fleeing in panic and the fraternity guys gathering outside with sticks, threatening to go in there and "teach those mutts a lesson." Eventually, just like in the monster movies, heroes arrived (grad students from the Vet School armed with tranquilizer darts) and, assisted by the same campus police who had worked the centerfold case, calmed the dogs with some well-placed shots. We spent the rest of the day cleaning the place up and getting the slobber stains out of the carpet. A couple of days later somebody let the same dogs back in. It was tense for a few moments but they just flopped down on the floor for a nap, as quiet and well behaved as freshmen. The affair made all the newspapers.

By my junior year I had "seniority" and was assigned exclusively to the Main Information Desk, the bridge of the Starship Enterprise. The pay was up to $1.60 per hour and I worked 30-40 hours per week. It was a great place to meet girls. I knew every co-ed in the arts college. They would call my name seductively as they approached the Desk. They knew I had the latest Reserve Reading List for any course on campus.

I have been thinking about why I didn't become a librarian, forget the Ph.D. and go after a MLS. I think it was the lousy pay. Plus, I fidget too much. But I have a small library in my home, just across from my office. I don't collect books and the shelves are mostly filled with old college paperbacks and monographs purchased during graduate school. The "periodicals" section is on the coffee table. In a small way it recaptures the atmosphere of a library and I can sit there undisturbed and think. I've noticed that when I have a tough business decision or tense phone call to make, I do it in the library where I feel comfortable and in control.

When I left college the information management skills I had acquired helped me get and hold a good private sector position with a Fortune 500 company. My initial assignment was in technical publications and training. I spent my first year gathering data from the company's archives, researching, organizing, setting up filing systems, classifying types of information...all things I had seen done a thousand times before at a library. My boss thought it was great; the information vital to product support was organized at last! Soon I had a staff of people working for me, accessing this data and writing, publishing and distributing technical and training information on all of our product lines. Every report was a term paper, every manual a thesis. Best of all for the company, the system I set up, like any good library, lasted long after I had moved on to other things.

The Digital Challenge Ahead

A few months ago John Meador, Director of Binghamton University Libraries, came to Rutledge, GA. Together with Laurie Miller, Director of Library Development; we spent an afternoon talking about the library system at Binghamton and about university libraries in general. They invited me to the campus for a visit and to meet with students and library employees and administrators. It was a great trip and it rekindled my interest in university libraries. As a result, I spent some time familiarizing myself with current issues effecting libraries.

What I learned is that the world in which libraries operate has changed dramatically in the past few years and that more changes are in store. In fact, we have entered whole new era for libraries. This is one of the most interesting and challenging periods in library history.

Let's survey the landscape. There are about 14.5 *million* college-age students in the U.S., about 5% of the total population. There are about 4,000 academic libraries in the United States. About 2250 are four-year (and up) institutions and about 1750 are two-year schools. Currently, approximately 80% of all U.S. colleges and universities (and their libraries) are public institutions supported primarily by taxpayers. Nobody is quite sure how many libraries there are not attached to an institution of higher learning. This is because of the great variation in size and type. Some, like the Philadelphia Free Library with 55 locations and 10 *million* items, are huge and command professional staffs and budgets that exceed all but the largest universities. The American Library Association (ALA) estimates that public libraries not associated with an academic institution received 1.2 *billion* visitors in 2004 and respond to 7 *million* reference desk inquiries *per week*. But many, like the one in my hometown, are tiny, all volunteer and operate on minuscule amounts of money coaxed year-by-year from local donors and town and village governments. And don't forget libraries (often called "media centers") at the secondary level. There are probably thousands, some good, some neglected.

And what are the trends in the use of libraries by students and the population at large? Here the statistics clearly define the challenge at hand. In September of 2002 the Pew Internet & American Life Project released the results of an extensive survey entitled *The Internet Goes to College: How Students Are Living in the Future With Today's Technology*. It showed that 73% of all college students surveyed were using the Internet more than the library for "information searching." Another 16% reported using the two (Internet and library) about the same amount of time. In other words, 89% of all students used the Internet as much or more for research. And while the survey showed that students still use the library, 80% reported using it less than 3 hours *each week*.

The survey results get even worse. According the Pew researchers,

> "During direct observations of college students' use of the Internet in a library and in campus computer labs, it was noted that the majority of students' time was not spent using the library resources online. Rather, email use, instant messaging and web surfing dominated students' computer activity in the library. Almost every student was observed checking his or her email while in the computer labs, but few were observed surfing university-based or library web sites. Those students who were using the computer lab to do academic-related work made use of commercial search engines rather than university and library sites."

And it gets *even worse*. One of the bright spots for student use of libraries is thought to be "study groups" wherein students go to the library and, using computer resources, collaboratively work on class assignments. Faculty members have assigned an increasing number of group projects in recent years. But when the Pew researchers observed these study groups they found (through direct observation)

> "...it is common for students to gather in groups and work in a computer lab for a prolonged time. While in groups, students often appear to be working on academic tasks although most often one student is at a

computer terminal typing *while the remaining group members are socializing* and contributing information when asked by the typist."

Conclusion: because of the Internet, libraries are experiencing a utilization meltdown of unparalleled magnitude. This is a potential catastrophe.

Are libraries aware that this is happening? For the most part, yes. After visiting a few, my sense is that libraries today are rethinking their place on the campus of the 21st century. If they are not, they should be, because they are in danger of being marginalized by rapid, radical transformation both of information technology and popular (student) culture. As compared with when I was a student, some libraries have started to look, feel act more like museums, stately, respected, but staffed by curators and docents rather than cutting edge "information acquisition facilitators." It is interesting that, at least where I have visited recently, the librarians seem disproportionately proud of the "special collections" that they have accumulated, and have earmarked a lot of scarce resources for their preservation. Their prized possessions are static displays of a sort you might expect to find at a museum, painstakingly assembled but rarely used by students or Faculty. Indeed, "protection" and "preservation" seem to be the current watchwords even for digital activities, which are focusing heavily on art and images as opposed to books and text documents. (Preservation activities may also be preferred because they offer a safe haven from copyright issues.) If this trend were to continue, library functions could be reduced to preserving marginal or exotic information useful only to a narrow range of researchers and accessible mostly by means of antiquated, manual methods of storage and retrieval. Greater reliance on technology and perhaps new institutions altogether would be needed to supply information to students and Faculty interested in subjects more in the mainstream of study and research. The library itself could become far less relevant to daily educational activities.

Do libraries that give priority to preservation and the status quo do the best job of supporting on-campus undergraduate education or community learning? I for one don't think so. To support undergraduate studies or communities, libraries need to be easily accessible data stores of current, vital information, central and not peripheral to campus and community intellectual and cultural life. Simply referring students to Internet search engines will not do. Libraries need to develop and then aggressively promote a complete array of academically sound study resources accessible using the latest technology, and they need to train (both directly and by example) today's already digitally smart students to use these resources to the fullest extent possible. This is absolutely essential if graduates are to step out into the real world of information intensive 21st century commerce and communications. Nostalgia aside,

I think that if static preservation of information were to become the predominant theme of library activities, especially for institutions with a large undergraduate population, it would be a disaster for American education.

The Future of Libraries

Many libraries are aware of the challenges they face and are mobilizing to respond. Almost everybody has some sort of "initiative" underway, most with the word "digital" or "electronic" prominently featured. There is Harvard's *Library Digital Initiative* (LDI), Cornell's *Digital Initiatives at the Library* web site, Columbia's *Electronic Publishing Initiative*, the University of Chicago's *Electronic Full Text Sources*, Berkeley's *Electronic Resources*, and so on. Some of these programs show great imagination and potential. Others seem more for show. Content is often added at a glacial pace. The Gutenberg-e project, for example, a prominent joint venture between Columbia University and the American Historical Association, claims to have 12,000 books online available for free download but added less than two dozen books in the first eight months of 2004. But some programs are more ambitious. The New York Public Library has announced plans to make all public domain content in their system, millions of volumes, available online at no cost to users over the next few years.

So, there seems to be a consensus among professional librarians and education experts that major changes are occurring in information culture and technology, but there is also widespread disappointment at the lack of substantive response. A late 2004 update to the Pew Internet & American Life Project *Report on the Future of the Internet*, which surveyed 1286 "experts" on this subject, reported that fully 50% of respondents were

> "…startled that educational institutions have changed so little, despite widespread expectation a decade ago that schools would be quick to embrace change."

Why is this? Everyone agrees that significant changes are occurring and meaningful response is overdue, but where is the "sense of urgency" typical of a business, institution or profession under stress? I find this both puzzling and troubling so I asked a sampling of administrators, librarians, Faculty and staff about it. Generally speaking, their responses fall into two categories.

It's Not a Problem, At Least Not Here

First of all, there seems to be a widespread *failure to recognize the extent to which digital technology affects their students*. The highest-level administrators understand the sudden impact of technology and are anxious to respond. But overall, the affect technology is having is substantially underestimated. The feeling is that, while the millennials are out there somewhere (we will discuss exactly what we mean by millennials in Chapter 7 on "Digital Students,") there are very few at *their* institution. The evidence offered in support of this view is interesting. Some, at what could be described as the more elite schools, think their students are "too smart and mature" to waste their time downloading music or blogging. They see their students as serious, focused, and career-oriented, rising above the "culture of the streets" where undisciplined youth wonder aimlessly into a cyber-world of illegal file sharing and gaming. Others, at less selective or open admissions institutions, think the opposite. They view their students as too poor to have regular access to technology. Therefore, they are not as digitally savvy as kids from affluent families, whom they see as more likely to web surf or send text messages. Both camps agree that "boredom" is a possible factor leading to preoccupation with new electronic gadgets. Faculty members say they don't notice a lot of students with cell phones or see them coming to class and taking notes with laptops. (They may forget that the tapping sound of keystrokes during lectures or the use of cell phones between classes is usually prohibited because they are thought to be disruptive to the learning process. Digital devices are also prohibited due to the possibility of cheating on tests.) The library staff reports that students use the computer terminals for only a few minutes and then turn to traditional textbooks or periodicals. Either that or they leave the library and head back to the dorm. Digital technology doesn't seem to have a big effect on their studies in the library. (Pew survey results support this observation but draw completely different conclusions.)

Some administrators acknowledge that they have lots of millennials on campus but feel that their school provides adequate technology support and training. They feel that that they are on top of any communications issues with digitally literate users of library services. They cite as evidence the availability of online tutorials and "wheat from chaff" information seminars, usually aimed at freshmen and/or new users. None of the professionals seem totally comfortable with their programs, however, and worry that their training is dated. My review of some online tutorials indicates that some are updated often while others go years with no new materials added. Many are nothing more than lists of online search engines.

While these observations may have some validity, those who conclude that the digital revolution has not impacted their campus are almost certainly wrong. All evidence — local, regional, national and global — indicates that the vast majority students, regardless of socio-economic background, have become heavy social and academic users of digital information and related technologies. The fact that a large number of academic professionals do not see this suggests a serious disconnect between them and their students. Increasingly, professional staff and students no longer seem to share an information culture. If this separation becomes too great, the institution could be in serious trouble. As I conclude in Chapter 7, a university can fail for the same reasons a business can fail…by losing touch with its "customers."

There is No Money in the Budget

The second reason given for not responding more aggressively to the digital challenge is lack of money. Everybody I know in the non-profit world believes in the gospel of "new money," namely, that unless additional funding is available (defined loosely as money above and beyond the current budget,) nothing happens. As I discuss in Chapter 8 "The Library as a Service Business," in an era of relentless downward pressure on traditional budgets and a trend toward creative, even radical financing of colleges and universities, new money is increasingly scarce. Therefore, focusing on maintaining the status quo seems to be the best, safest and most affordable strategy.

As a private sector person, I have always been puzzled by the preoccupation of academics and non-profits with their budgets. When I discuss education with people I know in the field, the budget (and the hierarchies of decision-making that go with it) are usually the main topics of conversation. This is unfortunate because the endless quest for more money is self-limiting. Recently it has becomes defeatist and an excuse for inaction.

This is not the case in business. I converted my business from analog to digital format in the late 1990's and I developed and carefully tracked budget figures during the first year because of the uncertainty involved. But by the second year *the savings from productivity gains alone more than offset any increase in expenditures.* In fact, expenditures actually decreased because of the rapid decline in the cost of hardware and software, a trend that, if anything, has accelerated in recent years.

Conversion to digital services or to better service in general is not a question of the budget. It is a question of managing effectively so as to maximize productivity and then properly re-deploying productivity gains in order to implement new programs and services. Some things such as old programs are "lost" but much more is gained. This is the budgetary strategy of a growing and expanding service business. Instead of fixating on the confines and limitations imposed by a budget, a business measures the return on its expenditures by either;

(1) Providing the same output or throughput at lower overall cost, or

(2) Increasing output or throughput at a fixed level of expense.

Either way, both the business and its customers win, the business because of greater returns created by increased business and/or decreased expense, the customer by an expanding array of services available, often at lower market cost.

In the past two decades a powerful new suite of tools — digital technology — has been developed for creating and managing information. These tools have the capability, when properly configured and deployed, to simultaneously increase quality and throughput while lowering unit cost. As I point out in a section in Chapter 8 entitled "Better Accuracy — Greater Productivity," the impact of digital technology on productivity is not always immediately obvious to many people, no matter how professionally experienced or highly trained they may be. But it is the single, most important factor in the unprecedented improvement in U.S. productivity over the past decade. Economists are still trying to sort out all the lessons learned from this productivity boom (the power of irrational exuberance or the unpredictable nature of job growth, to name just two) but leading businesses have already embraced the most important ones. These include quantum leaps in output and quality combined with stable or declining operating costs. Businesses that have adapted have prospered. Many of those who have not, have failed.

I recognize that the labor-intensive nature of education (the desirability of low student/teacher ratios, for example) may limit the potential for productivity gains at academic institutions. This, in turn, necessitates bigger budgets in order to increase outputs or initiate new projects. But *a library is a unique organization within a university or community.* It provides information and the services needed to deliver it. Much of the new digital technology is specifically focused on what a library does. It enables dramatic productivity improvements in creating, storing, retrieving and delivering information, so that this technology can have a significant, positive effect on how a library fulfills it mission. Improving the productivity of existing operations frees resources for new projects. This enables the library to grow, experiment and change even when budgets are tight. Teamwork between imaginative, driven leadership and knowledgeable, enthusiastic employees can accomplish as much or more than new money. Any library that adopts this attitude can become a model of advanced information services. This is a realistic, achievable goal for all libraries and librarians. It should be written into the mission statement of every library.

Success Strategies

Unfortunately, there is no simple formula. Each library confronts different problems with a unique arsenal of resources. Each is inspired but also entrapped by its own history and traditions. But there are some general strategies that can be used by all. In this book I propose two,

(1.) Libraries urgently need to adapt to the culture and technology of digital information with the goal of delivering world-class information services to students, Faculty and patrons. Otherwise, the disconnect I believe exists between students and educators will become permanent.

(2.) Libraries can achieve this level of service by borrowing from the strategies and tactics of successful information service businesses.

Not coincidentally, the past two decades of my professional career have been dedicated to starting and growing information service businesses and I have managed the transition from all-analog to all-digital information processing. I think I have some useful insights to offer.

This book divides into two parts. The first six Chapters focus on business strategies and technologies that make an information service business successful. I start by providing some insights into how successful services businesses respond to change. A key is the aggressive introduction information content into their business model in order to attract and hold customers by adding value to their products and services. I then consider recent developments in the evolution of information services. The most significant breakthrough in this area has been the Internet search engine, which has had a huge impact on the availability of information (and which, since about 2001, has started to replace libraries as the primary source of academic information for students.) Search engines have worked to perfect a business profit model based on revenues from online advertising. Advertising has had significant effects, some negative, on the quality and content of services provided. I also devote a Chapter to other unresolved problems with Internet search engines that limit their usefulness to users. In a separate Chapter I explore issues relating to popular music that are playing a key role in the evolution of online digital technology and culture. In the early Chapters I also explore two of the most powerful distribution technologies in our future information environment, peer-to-peer (P2P) technology, which so far has been used primarily for music file sharing but has vast potential beyond that, and wireless Internet access (Wi-Fi) which is struggling with issues of public policy that have, to date, slowed deployment. Obstacles remain to the universal acceptance and availability of P2P and Wi-Fi, but it seems only a matter of time (and consistent public policy) before they enable all information to become Ambient Information—available anywhere, any time and in almost any imaginable format.

The second part of this book, Chapters 7 through 10, focuses on things a library must do to become a successful service business in the digital age. This is more about integrating people with technology than building a business model. I propose that libraries, using the example of private business, recognize that the most important thing about any information services business is the customer. Students—called millennials or thumbs—are part of a rapidly changing digital culture. This culture is already dominant among most high school and college students. Libraries must learn to do what any information business would do when threatened by change, that is, structure their workflow and hire and train staff to synchronize with this unique student (customer) culture and the technology that supports it. Applying business strategies to library services, including progressive hiring profiles and aggressive marketing strategies that draw attention to unique programs, can help create superior, even world-class, undergraduate teaching and research support for students and Faculty. This approach also has the potential to improve the quality of information and overall productivity within libraries, reducing budgetary pressures while at the same time enabling them to compete more effectively with the for-profit model of digital search companies. In order to avoid becoming analog museums, libraries must provide services that offer a viable alternative to Internet search engines and other commercially driven information sources. Otherwise, the library-centered academic campus community, which for centuries has made university or college attendance a unique, life-defining experience, will suffer irreparable loss.

Finally, there is a concluding Chapter in which I recommend some steps that every library can take to reverse the disconnect that is occurring between traditional library services and the millennial generation. I intentionally focus on changes that require little or no new funding but which can have significant short-term impact as part of a long-term response strategy to the digital challenges at hand.

I would like to make one additional point about this book, that regarding style. I have not gone back through and edited the language and deleted the personal anecdotes to make it sound more "scholarly," nor have I cluttered it with footnotes, bibliographical references or other academic paraphernalia (the exception being a detailed Index.) Also, I have omitted most of the library and technology jargon. Having waded through some dense academic and technical papers and monographs as part of my background reading on this subject, I can vouch for the fact that they can be difficult to absorb except by the most hard-core library or technology professionals. I supply some definitions when necessary and place some of the less familiar terminology in quotes so readers can pause for a second and intuitively decipher new vocabulary and concepts. (Also, without footnotes, I have been forced to use parenthetical expressions for asides.) All of this may seem inappropriate for a book primarily intended for readers with an academic background. But I have tried to write something that is a little more casual so that it might also be of interest to a broader spectrum of alumni and friends of libraries and higher education as well as university and library administrators and employees, students and Faculty members. Perhaps by reading this, leading elements from within these groups will develop a greater interest in the future of libraries and information in general. That is my hope. So, it would be a mistake to conclude that the work is academically or intellectually sloppy based on style alone. The anecdotes may strike you as sappy and some of the ideas a bit outside of the box, but you can have confidence in the facts, figures, quotes and other references I have used. They have been properly researched and they are accurate. After all, I have learned from librarians.

2. The Development of Information Services

The Digital Economy

In recent decades, "service" has become a major part of the U.S. economy. U.S. manufacturing companies have declined in number or, more recently, moved offshore, but service businesses have multiplied to the point where we routinely speak of a "service economy" and a population of "service workers." Within this service economy differentiation has occurred. Service no longer refers just to lawn maintenance or fixing things that break. It also includes the assembling and distribution of information, a far more complex task. Despite or perhaps because of this complexity, the "information economy" now dominates and defines the service economy. A business today that produces information has a far better chance of prospering than one that merely consumes it. The future of wealth accumulation is information services, and only secondarily non-information intensive services such as machine repair or cutting hair.

How did this come about? What happened to creating wealth the old fashioned way, through farming or manufacture? In the past two decades three things have occurred. First of all, *the value of information has become more readily recognized.* The reasons for this are many, including the development of more sophisticated products and services and the business models to support them, the stimulative pressures of global competition, the universality of broadcast communications, greater governmental regulation, the global trend toward democracy, a better educated population, and some fortuitous discoveries (CRTs, transistors and silicon chips, to name a few.) Producers see the competitive advantages that are possible with more and better information; buyers enjoy greater efficiencies of consumption. The word both in the Boardroom and on the street is "you snooze, you lose." Translation: what you don't know can be highly detrimental to your business or your personal well-being.

Second, since the mid-90's, *the Internet has exponentially increased the opportunity to distribute (and access) information.* There is now a global information pipeline directly into every home and business that has wired telephone service. Even wires are becoming unnecessary as wireless telecom spreads. The initial cost of household or business access is already approaching zero and the marginal cost to both information distributors and consumers is absolute zero. Not surprisingly, business has rushed in to take advantage of this, further heightening our awareness of information and its potential impact on our lives. Internet-based information businesses have become wildly growth oriented and brutally competitive in only a few short years, even before most of the players have developed a viable business model. Businesses compete to make information more and more useful. Thanks to their efforts, the promise of an unlimited supply of infinitely variable information, customized per individual, seems well within reach.

Third, *the world is undergoing a radical transformation of popular information culture.* In just twenty or so years hundreds of millions of people around the planet have already learned to transfer information digitally, using standardized, simple-to-learn graphical computer interface. Neither language, nor geography nor even lack of formal education is a barrier. Once this communications skill is mastered (and mastery comes relatively quickly for most people,) knowledge, information and entertainment can be accessed or even created by any person and shared with thousands or even millions of other people, instantly. This has been facilitated by the proliferation of inexpensive, universally compatible hardware (primarily the personal computer — PC.) This has changed the way the "digitally literate" among us gather and use information. Increasingly, they are coming to expect all information to be "ambient," that is, personalized and available digitally everywhere at any time. (I discuss ambience in detail in Chapter 6.) Information that is digitally inaccessible or otherwise constrained loses its usefulness. So far, this phenomenon has primarily affected students and young professionals, but as they pass their culture on to their offspring, the whole of global society will be transformed.

Where is this leading? Unfortunately, the new information culture has the potential to inflict severe collateral damage to some of our best, most cherished information resources, especially public and academic libraries. That this damage is for the most part unintended does not diminish its effect. Not all libraries are located at universities, but all colleges and universities have libraries, so that the challenge to libraries is also a challenge to the campus educational experience to which most students still aspire and which campuses still try to deliver. Can libraries (and their universities) adapt to the new technology and remain viable and competitive? This is the challenge ahead.

The Inner Dynamics of a Service Business

Since libraries are generally non-profit organizations, information business profit models (which themselves have many inherent problems) are not helpful. But, other elements of the private sector service business models hold great promise. Let's look more closely at small-scale private enterprise. A successful for-profit service business has *an inner dynamic that may not be obvious to outsiders.* Nonetheless, this dynamic determines the level of success the business will ultimately achieve. There are three characteristics that are especially important,

- A sense of urgency
- A willingness to take risks
- An understanding of what kind of services satisfies customer need.

A Sense of Urgency

First, a successful business has a sense of urgency in responding to changes and challenges (consumer tastes, for example, or advances in technology.) Owners and managers must develop a business structure that accommodates rapid change. Forward thinking owners worry constantly about the speed with which the business is reacting to new opportunities, products, services, technologies and competitors. Change becomes an ongoing activity, firmly imbedded in the organization's culture. Ultimately, a really aggressive business creates new opportunities for itself by seeking change even when there appears to be no changes looming on the horizon. Changes may be implemented in part just for the sake of changing. The challenge of change gives success-driven businesses a powerful adrenalin "rush." Change becomes proactive, not reactive. The business gains a reputation among both employees and customers of being "aggressive." Product selection and service constantly change and improve. Customer loyalty grows; brand identification builds; employees are energized (otherwise they are replaced.) Competitors are now the ones who are constantly worrying.

How to Lose Business by Laughing at Customers When They Need Your Help

Here is a quick way to measure a service organization's sense of urgency. Go to their office or retail store. Look around and see what they have taped to the wall, front counter, credit card terminal, etc. Do you see that ubiquitous cartoon with the little nebbishes rolling on the floor laughing and the caption above saying "You want it WHEN?"...Or how about the one that ridicules the customer who is behind schedule and needs immediate help "Lack of planning on your part,...?" These cartoons are cute, but they reveal an organizational mindset that ridicules urgency. This is being communicated directly to customers and fellow employees through posted signs that, apparently, are approved by management (otherwise management would tear them down.) This is the wrong message for both customers and employees and the business suffers as a result.

A Willingness to Take Risks

In addition to a sense of urgency, a successful small business has a tolerance for risk-taking that is high enough to enable frequent change to occur. For our purposes we will define risk as making decisions without enough information to be certain of the outcome. Another way of describing risk-takers is to say that they have "a high tolerance for ambiguity." Risk does not mean crazy, gambling risk. The most successful business people I know are risk-averse. They want predictable results from their actions with *no unintended consequences*. This rarely happens, however, because the speed and frequency of decisions that are made in a small business, which does not have the market or technical research resources that a larger corporation may have, preclude having all the facts available at decision-time. Limited availability of information means an imperfect decision-making process. The urgency with which an aggressive business responds to situations and a tendency toward risk-aversion, both present in successful businesses, naturally conflict with each other. It takes skillful management to balance the two.

Service businesses that are successful display more willingness to move their businesses in a new direction even when the exact outcome is uncertain and risk is greater than they would like it to be. I think the reason for this is that a successful service business is more closely attuned to the needs of its customers because it is selling service, not a generic product being manufactured and packaged at a remote location. The service business, unlike manufacturing, seeks to cultivate repeat business and receives constant, real-time feedback from customers. Satisfaction is measured continuously. A successful service business learns, sometimes only by trial and error, that insensitivity to changes in customer need can result in loss of customer loyalty. This is one reason why small service businesses can often outperform big service businesses; the owners and managers have a closer relationship with their customers and receive (if they are listening) unfiltered feedback every day. This is a tremendous advantage in a competitive marketplace or in a period of rapid change. If the business is willing to take risks, immediate feedback results in a better understanding of customer need and a quicker decision-making process. (This is also a reason why the failure rate of small business is higher. They make more mistakes.)

Fear of change is the enemy. Once a business stops pursuing change (or if it fails to start,) management tends to continue along that path because avoidance of change becomes the dominant mindset within the company. The effects of not embracing change multiply, increasing the bias toward inaction. The "entrepreneurial spirit" is lost (or never unleashed to begin with) and the internal business dynamic shifts toward inaction. This is a common reason for the failure of a small business; it enjoys a successful startup, but then, having failed to respond to new opportunities and challenges, it dies.

Customers are often the big winners when businesses take risks. Customers gain access to new products and service, receive faster service, enjoy a better retail experience, are assisted by more knowledgeable and energetic employees, etc. Typically customers respond by using more of the new products and services being offered. Their satisfaction and loyalty grow, and the business prospers.

A Case of Unintended Consequences

When initiating change, there is always the possibility of unintended consequences. This is a normal part of risk-taking. We can define unintended consequences as results or events that were not anticipated when the original project or action was conceived. They can be either good or bad. They can occur whether or not the original project is successful. They can be short-or long-term. All of these possibilities heighten the risk factor in project planning. Sometimes these consequences are so powerful that there is no way to control them once they begin to unfold. An extreme, but true, example will illustrate this point.

A couple of years ago a major northeastern university decided that they would videotape some lectures and make them available library-like to students who had missed the class due to excused absence. This sounded reasonable and the Faculty agreed. However, nobody anticipated the results of these videos being made available over the Internet, as they soon were, formatted as streaming multimedia files that could be played on any laptop by applications such as Real Networks or Microsoft Windows Media Player. Soon many students began watching these videos even though they could have attended the real lecture. It was just easier to stay in the dorm or apartment and view multimedia on their PC's (which they were doing a lot of anyway). The university embraced the idea and began providing links to the online multimedia lectures (and to the necessary client software) from the official university web site. This has become an increasingly popular program at that institution.

The changes this could bring to the campus are far-reaching. Sooner or later, all university lecture courses could be available this way, and not just for enrolled students but also to anybody, anywhere, who was interested. This could extend to students to other colleges and universities. Password protection might appear to limit access to tuition-payers, but outlaw media companies could hire students to bootleg lectures for black market resale (the same way they record rock concerts) by sneaking video cameras into the lecture hall in hollowed-out books. Would students buy these unauthorized videos? Well, complete term papers are regularly bought and sold on every campus, fans buy bootlegged concert tapes and students illegally download music files without too much hesitation, so why not lectures?

To protect their intellectual property rights, the university and Faculty could begin restricting access to lectures, allowing only carefully screened students into the "audience." (The audience could also be coached to laugh or applaud on cue, just like TV audiences.) Why use an audience at all? Multimedia lectures could be created in a studio, where audio and video facilities are much better than in a lecture hall or classroom and authentic background sets could create the right setting. The Faculty could syndicate their lectures to multiple universities, serving on the research faculty of only a couple but "teaching" at ten or twenty each semester. Of course, to become true "Professtars" they would need to polish their "edutainment" persona and add an array of multimedia (film, music, animation, etc.) to their lectures. Every Professor in the Psychology Department could strive to be as smooth as Dr. Phil. The English faculty could emulate Oprah's book club and, as a reward for signing up for the course, give each student a car.

Universities and Faculties nationwide would join forces to pressure Congress to change intellectual property laws in order to bring education more in line with the rights enjoyed by entertainers. The definition of "fair usage" could be narrowed to prevent personal duplicating or even passively watching a lecture without paying license fees or tuition. Copyright would be extended to protect anything a Faculty member might say, including impromptu or extemporaneous comments made during office hours or over coffee with students (which might otherwise be recorded and illegally posted on some web site.) Meanwhile, the students would put their TIVO skills to work to reduce lectures to their essential core, "fast forwarding" over the rest. The part where the Professor wandered off-topic for a few minutes, or the boring question from the person sitting in the front row could simply be skipped. A one-hour lecture, like a one-hour TV show, could probably be cut to 30 or 40 minutes. As for the Faculty, syndication would lead inevitably to TV's most endearing and enduring product—the rerun. Once a history prof has videoed a brilliant lecture on the complex reasons for the decline of Athenian democracy, why ever give it again? Just play the tape over each year.

Students, Faculty, the university administration and even alumni might think they are the long-term winners in all this. Students would get better edutainment but spend far less time in class. The distinction between students and alumni would be blurred, as alums remained perpetually enrolled in courses long after "graduation" while never actually setting foot in a classroom again. The Faculty might see a complete transformation of teaching as we know it together with an extraordinary flowering of academic research stemming from the decline of in-class teaching responsibilities. The university's construction budgets could be cut because fewer classrooms would be needed. On the other hand, more teaching assistants (TA's) would have to be hired to grade papers and hold labs and traditional "discussion sessions" for any students still wanting to actually talk about course material. Top Professtars might cause enrollments to skyrocket by attracting thousands of distance enrollees, who could buy a popular course for the price of credit hour tuition. But a campus that had a sense of itself as an intellectual community might be diminished as more and more academic discourse occurred electronically.

To summarize, the lecture video program has now taken on a life of its own, one that is substantially different than what was originally expected. Did anybody think this program through before it was launched? Could the university have anticipated what was to happen? Should they have risked letting the program go forward without knowing what might happen? Or is the control of technology, once unleashed, impossible and is this merely a glimpse of an inevitable future, one that improves the educational process and benefits Faculty and students alike but changes forever the traditional campus learning experience?

Understanding What Kind of Services Satisfy Customer Need

If a business is a service business, it needs to have a clear idea of *what* it is offering its customers and *why* it is of value. Products are easier to define and compare; service is less precise. But a successful service provider struggles with this issue and eventually comes to understand what satisfies customer needs. Otherwise, the business fails.

Every customer wants and at times needs service. But what is service? Unlike, "risk," there is no short definition. Service is intangible and often is defined entirely by customer *perception*. Therefore, various kinds (or levels) of service need to be understood in the appropriate context. An organization has to decide what kind of service it should offer to best meet the needs of customers while at the same satisfying the metrics of its own business model.

Standardized Service: No Information Content

Let's consider "service" on your dishwasher. This is usually a one-size-fits-all program the purpose of which is to get the dishwasher back up and running at a reasonable cost in a reasonable amount of time. For any specific machine model, almost everything about this kind of service is generic. The functionality of the appliance is the same in every home (clean the dishes.) For any given problem, the parts replaced are the same (if the water pump fails, they replace the water pump plus associated hoses, gaskets, etc.) Most parts are prepackaged in "kits" at a fixed price. The labor component (skills plus time) is the same for every customer and is usually priced at a "flat-rate" for all customers. In fact, service of this sort (parts plus labor) is usually priced at a "contract" price that is pretty much the same in California as it is in New York. Customers are urged to pre-purchase service at this fixed rate (contract price.) They may even be promised "priority" response time if they pre-pay.

The sum total of this type of service is called a "service contract." The salient feature of a service contract is a built-in comfort level for both the customer and the service provider. The customer knows in advance how much a repair will cost and approximately how long it will take. The service provider knows in advance how much time will be required and how much revenue will be generated by this service (and also, if they keep good records, approximately how many water pumps will be needed for a known number of appliances sold over a given period of time. Parts inventories can be adjusted accordingly.) Because of increased predictability of cash flow and expenses combined with management, logistics and training efficiencies, service contracts can be hugely profitable for a business. And because of the repetitiveness of the repairs (which can become rote) employees also favor predictable processes that, once mastered, are low-stress and personally satisfying. Everybody is happy.

Literally thousands of products, from automobiles to xenon lighting systems, are repaired and maintained through standardized service contracts. Repetitive maintenance, such as lawn or hair care, can be placed in the same general category. It is a large (and usually growing) part of the total US economy and the single largest profit contributor for hundreds of businesses worldwide, including such well-known names as IBM, General Electric, ChemLawn and Great Clips. Because of this track record of success and profitability, the idea of standardized service has been distilled into a business model that is emulated by hundreds of thousands of startup companies each year. This is an important component in our so-called "service economy."

Differentiating Standardized Service With Information Content

In recent years, many different types of businesses have found themselves competing in a marketplace overcrowded with generic or standardized products or services. In response, they have attempted to differentiate themselves by *adding information* to increase the value of the product or service they provide. The goal is to turn something generic into something with unique appeal to potential customers because of its information content. This is not always easy to do but there have been some imaginative strategies developed to create an information component for various generic products or services. Quick printers, for example, not only have been able to establish the word "quick" as part of their name (implying superior service due to fast turnaround on print jobs) but also have been able to redefine printed materials as "information." Cutting-edge print shops, whether independent or franchised, now sell information that may be put on paper but just as often is digitally "printed" onto CD's, disks or is posted on web sites for customers to view. Printers who have been able to capture information, format it and then distribute it on either printed or digital media have prospered. Those who have not have failed. Industry experts estimate that as many as one-half of all "quick print" businesses have closed in the past few years. They discovered that "quick" alone is not enough. There also has to be useful information to build customer loyalty. The alternative is unrelenting price competition for a service (printing or copying) that from the customer's perspective is generic.

Another example of an industry burdened with generic products is the "fast" food industry, which has built hundreds of thousands of outlets and has struggled with market saturation in the past decade. The word "fast" is used generically by the industry itself to describe the level of food service provided, although in the early days of the industry only certain of the hamburger chains claimed to be "fast." McDonald's, for example, used to promise meal service in a specified time, usually 2 minutes. McDonald's would place small timers (clocks) on the front counter that would be started when the customer placed the order. If the food was not delivered by the time the bell rang, it was "free." [Quick printers also used to use the same gimmick; the print order was either completed on-time or it was "free."] But when everybody is "fast" or "quick" this strategy loses its effectiveness. The appeal of fast service to the food chains has also diminished because it is generally more costly.

Like printers, the food industry has begun to discover the advantages of spreading information as well as mayonnaise on the product being served. In some cases, the information is about the food itself. The various health and diet fads have prompted companies like Sub-Way to tout the "healthy" nature of their sandwiches as compared with hamburgers, fried chicken or French fries. Sub-Way prints actual comparisons of calories, fat and sodium content, etc., on their napkins and drink cups. The goal is to use information to build customer awareness of the health advantages of their products vs. the competition. Presumably this creates loyalty even if the service itself is not especially "fast." In fact, Subway does not promote superior service, only comparatively healthier food.

But many food products are simply not healthy (or at least their claims are doubtful) and have to develop even more imaginative strategies using information or entertainment (a form of information) to promote themselves. Coffee, a product that I have been known to enjoy on occasion, is a good example.

Why Are There So Darn Many Starbucks Coffee Shops and How Do They Get People to Pay That Much for a Cup of Coffee?

One day in the not-too-distant future there will be 10,000 Starbucks coffee shops in the U.S and another 15,000 overseas. This is a lot, but not a record number for a "food" chain. Many other companies, including McDonald's, Wendy's and Sub-Way, have more locations, at least in North America. What makes Starbucks unusual is that while the location of most fast food outlets roughly matches the distribution of the general population spread along main highway arteries, Starbucks seems to clump their stores together and even open new ones continuously in the same areas. It is not uncommon to be able to see another Starbucks from the one you are at, immediately across the street, for example. Multiple locations in the same mall are also becoming more common, again sometimes within sight of each other. Humorist Dave Barry says the Starbucks corporate motto is "There's one opening right now in your basement" (their actual motto is arguably better, "Sometimes the coffee stirs you.")

The explanation for the success of this business model is Starbucks' clear understanding of who their customers are (students and urban professionals) and what their customers want, which is a predictable level of *service*, not just good coffee. Coffee is not a nutrition food (like lunch or dinner.) It is an optional, luxury item that is as much about the "moment" as it is about the product. By moment I mean the fun of grabbing a quick cup on the way to the office in the morning or on the way back from an important meeting or after lunch. Sometimes it is just a case of getting out of the cubicle or office and away from the mouse for a few minutes to clear your head and do a little attitude readjustment. (The English achieve the same effect by drinking tea and turning tea-time into more of a relaxed, social occasion, but unfortunately for busy Americans it takes longer to brew, serve and savor tea than it does coffee.) But the amount of time the average customer has available for this mental getaway is brief, sometimes only a few minutes, as compared with 15 – 20 minutes for a "coffee break" or 30 – 60 minutes for lunch. So a lot of coffee customers are on a tight schedule. They want to know in advance just how long they will be delayed if they stop for a cup of coffee.

The time crunch is worsened by the fact that, due in part to competitive pressures, Starbucks has opted to differentiate its coffee into a wider variety of flavors and concoctions, all of which take even more time to prepare than regular coffees.

This means that the customer has to wait even longer. They have great tasting treats, like Vanilla Bean Frappuccino® Blended Crème or the seasonal Pumpkin Spice Latte, but who has the extra time?

Starbucks has chosen to respond to this problem in a counterintuitive way. First, they note the amount of time a customer has to wait to place his/ her order at the counter. When they see the time lengthening to the point that an "aggravation factor" becomes part of the service equation, they open another coffee kiosk close by, preferably within walking distance of the one that is overloaded with customers. Customers who are really pressed for time now have a choice of two shorter lines. As customers leave the first kiosk and buy their coffee at the new one, the original kiosk lines grow shorter and service improves. The new kiosk has less business and the wait is shorter there as well. This is a win/ win proposition for the customer, which is precisely the Starbuck goal. An added bonus for Starbucks is that customers are willing to pay a significantly higher price for a cup of coffee that tastes, in my experience, pretty much like everybody else's coffee. For this reason, Starbucks has been profitable for some time now despite ferocious competition from less expensive coffees.

Most small businesses (or business units like a Starbucks coffee shop) are pleased to see a line form at their sales counter. It seems to validate their business plan when people are so eager to get their product(s) that they are willing to stand in line. I have experienced this myself, a warm, fuzzy feeling that business is good and customers like us so much they are willing to queue up to hand us money. But Starbucks has learned (probably by trial and error) that their product (coffee) and what their customers want (predictably quick service) are one and the same. They structure their business model accordingly. They are willing to cannibalize business from a successful kiosk and redirect it elsewhere just to maintain their service response time. To quote their Corporate Mission Statement posted on their web site, one of their goals is to "Develop enthusiastically satisfied customers all of the time." Nothing does this like great service. Even the English have started drinking Starbucks coffee.

The Starbucks management team, successful though they have been, has not been immune to failure. They decided a couple of years ago that they could use the power of the Starbucks brand name to market all kinds of products to tens-of-thousands of already-satisfied customers. They talked about clothes, furniture, and home appliances, all part of a Starbucks "latte lifestyle" (my term, not theirs.) They also launched a lifestyle magazine called "Joe." It was all a huge failure. The products didn't sell, the magazine folded and their stock plummeted on Wall

Street. What happened? I think that they strayed too far from markets where *quick service* is the decisive factor in a purchasing decision. None of these add-on products were time sensitive and many were far too costly to be purchased on impulse. (High-end espresso machines are really expensive!) The same customer who loved Starbucks coffee, continued to purchase his/her lifestyle products from Frontgate or Sharper Image, who have greater selection at lower prices. It was a lesson in humility for Starbucks, a descent into "commodity hell."

Nor is the Starbucks service-focused business model immune to competitive pressures. Dunkin' Donuts, with 5,000 locations in the U.S., is installing a precision Swiss-made espresso machine that makes consistent *cappuccinos* in less than 60 seconds. While Starbucks has hires employees to manually create their drinks, Dunkin' Donuts says their service is "idiot proof" thanks to the new hardware. They price their drinks about 25% below Starbucks due to the labor savings. Customers who have converted to Dunkin' Donuts from Starbucks cite the faster service and greater predictability of their experience vs. Starbucks, so the competition may have found their own version of the Starbucks service formula.

So, despite a unique business model and a great deal of success serving coffee, Starbucks finds itself under intense competitive pressure. Starbucks management has realized that they may not be able to count on fancier coffee drinks to carry the day. Instead, they have turned to a new form of product enhancement, *information*. They recently purchased one of the smaller music download companies and are offering a completely new service at their coffee kiosks…music downloads onto a CD while the latte is being frothed. Customers can call ahead or order over the Internet and their CD will be ready when they get there. Or, they can "burn" the songs onto their own laptop at a listening station while they sip their espresso. The price is low, $.99 per song, and the service is fast, due to broadband download speeds. Starbucks is working to expand their music library to appeal to a broader spectrum of tastes. They (together with many other companies including Wal-Mart and Apple) think the paid-download business is the future of music (and, eventually, of movies as well) and they have figured out a way to bring their superior service ethic (as well as their 10,000 locations) into play. This is a interesting idea, which compliments both the Starbucks Wi-Fi Net access program launched a couple of years ago and the "Starbucks Hear Music" channel on XM Satellite Radio, which has more than 2 million subscribers. It will be interesting to see if the Starbucks' venture into digital entertainment is successful. One thing for sure, it brings a whole new meaning to the idea of stopping by for a cup of coffee.

To summarize, Starbucks has recognized the convergence of its product (coffee) with customer need (predictably quick service) and has developed a business strategy to capitalize. Despite having a relatively generic product and very intense competition, they have developed a unique rapport with thirty million customers per week who pay a premium price for a few minutes of relaxation. Starbucks has also begun to blend information (entertainment) with their coffee service, a unique way to differentiate their service from the competition. There is a lesson here for any service business, which should find Starbucks' success truly "stirring."

Information Service: Selectively Variable Content, Random Customer Access

The type of service provided by an appliance repair business (or a coffee shop) is based on calculated needs and standardized, repetitive procedures. But not all service needs and solutions are as predictable as dishwasher repair or coffee service. This is especially true if the service is data or information intensive. Information services are more complex because there are more variables, including variability in content and users, and sometimes both at the same time.

For example, some types of information services involve standardized format and procedures but provide variable content. A good example of this can be found in the financial services business (banking, brokerage, accounting, etc.) where each individual customer needs his/her own specific data but where the format for presentation of data and the type of data are pretty much the same from customer to customer (tax or GAAP rules or requirements of regulatory agencies such as the SEC usually dictate format.) Standardization provides the same benefits to the financial institution as it does to an appliance repair company, including predictable needs and standardized, repetitive procedures. But the information is never the same for any two customers. This *brings tremendous value to each customer*, who may become totally dependant on the service being provided, which also happens to be the goal of the financial service providing the data.

Unfortunately, financial services have developed a tendency squabble with their customers about how much, if anything, customers ought to pay for this information. Confrontations with customer service are all too common, as anyone who uses a bank can attest. As a result, customers switch services with surprising frequency, thereby undermining the dependency factor that the financial services work so hard to achieve through customized data. The business model thereby contradicts itself, attracting customers with customized information but driving them away with price and poor customer service.

News information is another interesting example of an information service that provides variable data. The news, sports, entertainment and weather change every day. The daily paper and network news shows have to change accordingly, although they may present this new information only once or twice each day. Twenty-four hour cable news (Fox, for example) and weather (Accuweather) as well as Internet news sites (CNN, MSNBC) can change continuously throughout the day, although many days there is minimal change in the stories being covered. Instead of changing the information (which itself is not controlled or produced by the information service and which may not actually be changing during any given period of time) the cable outlets change the news readers and commentators, who present the same information but with a different persona. Internet sites frequently retain the text of the story but change the online photo(s) every couple of hours to give the appearance of freshness. The actual data may not be changed much over an extended period if time. Some stories remain online for days.

News services target a random audience (whoever tunes in) with the same news product (collection of stories and analysis.) You may not want to hear about tomorrow's weather but you will anyway unless you change the channel. An inability to target specific information for specific individuals may actually drive viewers and readers away from cable and Internet news sites. This is not the effect they are trying to achieve.

News information service is not generally cited as a successful business model. Unlike companies that provide service for tangible products (like dishwashers,) news information services do not charge for access to their content; news is given away *free* to millions of customers. News is supported indirectly by advertising sales, but broadcasting ads is an inefficient way to generate revenue. Many news services, especially network news, are unprofitable. Moreover, technology is providing viewers with ever-improving means of ad avoidance (TIVO or popup blockers, to name just two) so that the effectiveness of general advertising to a random audience is constantly being degraded. To counter this, advertising companies try to place products in the program itself instead of having a separate section of the show set aside for ads. This may work for entertainment shows, where the actors can hold a can of Coke, but is much harder to imagine in a newscast (can a news anchor drink from a Coke can during the show?)

To counter lack of profitability, news services have tended to cuts costs by make their presentation more standardized and less variable, *i.e.*, they try to be like standardized service business models (information needs determined by the content provider instead of the customer, and standardized, repetitive procedures for delivery, all aimed to reduce expense and improve profitability.) This runs counter to current information service trends, and traditional news services have declined in user popularity in recent years.

Information Providers and Information Consumers

There are about 25 million businesses that file tax returns in the U.S. each year. Less than 5% have revenues over $1.0 million. Over one million new businesses are started in the U.S. every year. A substantial majority would describe themselves (and would also probably be described by others) as "service" businesses. The top twenty types of startups in any typical year in the late 1990's include cleaning services, automotive repair, consulting, beauty salon, computer service, designer services, lawn maintenance, marketing services, landscape contractor, painter, investment services, and communications consultant. Many others in the top 20 including restaurant, real estate and retail sales have a high service component although they are not technically classified as a service business. Most of these businesses are sole proprietorships, not corporations, and have few if any employees outside of family members. Many have no employees at all; there is just one person providing the service. These "businesses" are not really organized businesses in a corporate sense. An individual starts a business seeking self-employment at wage (profit) rates that are perhaps slightly better than they would receive if someone else employed them. The driving motivator is to "be your own boss."

What is most interesting is that virtually none of the top 20 startups provide much in the way of information as part of their service. Instead, they are net consumers of information (parts lists and service data for mechanics, stock research for investment counselors, etc.) Business failures among startups are very high. There are many reasons for this including inadequate capitalization, poor management skills, "burnout" among sole proprietors, etc. But there is also a correlation between business failures in the top 20 startup group and the amount of information they provide. The lower the information content of their service, the higher the rate of failure. Conversely, high information content improves (although it does not guarantee) the likelihood of success.

The implications seem fairly clear. Insofar as success in business is concerned, it is better to be a provider of information services than just a consumer of information.

To summarize, as the service economy has grown and the value of information has become more clearly recognized, aggressive businesses have learned to use information to differentiate their products and services from those of competitors. Information intensive business models, from bankers to printers to coffee shops, have provided their customers with increasingly sophisticated information strategies tending toward variability and personalization. Information content has become predictive of business success. It should come as no surprise, therefore, that the Internet Search Engine, that has no product other than information and that has learned how to offer infinitely variable and personalized content, has become one of the most successful business models in history.

3. Internet Search Engines

The most recent development in information services is what is called the Internet Search Engine. They attempt to provide variable information that is fully customizable for each individual. They also attempt to expand the range of available information to almost any subject. They do this by asking the searcher to provide input on what he/she is looking for by entering a customer-designed free-form query into the search bar. Information (which is limited to URL's — Uniform Resource Locators, — web site addresses) is then selected by the search engine from a variety of resources including indexes of web pages that have been prepared in advance (and updated continuously.) Searchers can then select and "visit" web sites by clicking on the URL provided. The decision as to which web sites to visit is, in theory at least, left up to the searcher, although search engines are learning how to guide user searches toward specific information (as we will see, this has become a key part of their business model.) As search engines have become more sophisticated they have expanded search into new areas, including images, video, music, TV programs, even shapes. The goal of search engines is to eventually provide search capability for all of the information available anywhere on the Internet.

The search engine phenomenon is revolutionizing information services by providing fully customized, infinitely variable information to each individual customer. Pundits are already declaring that this is the first paradigm shift in information distribution in the 21st century. The popularity of this type of information has exploded in just the past 2-3 years. A 2001 study, for example, revealed that 79% of all students were initiating information search by clicking on a search engine site instead of using more traditional sources such as the library. By late 2004, a follow-up survey revealed that student use of search engines is right at 100%, that 89% of all Internet users query search engines, 87% of those using search engines are satisfied with the results they get, and that 35% of all users, approximately 38 *million* people, consult search engines daily. (Although all of this sounds impressive, the survey also uncovered some huge problems with search engines, which are discussed later in this Chapter and the next.) Although search engines are a relatively new business (Google, for the moment the acknowledged leader, was started in 1999 and didn't become a public corporation until 2004) there are several powerful competitors, including Yahoo, Microsoft, Ask Jeeves, Amazon A9 and AOL. Less well-known competitors (Mooter, Kartu, Vivisimo, Snap, blinkx, and many, many others) are emerging, as the technology is refined. Because of the rapidly growing and apparently limitless demand from customers for this type of service-individually customized information at little or no cost to the consumer—enormous effort has been made to develop not just better technology but also viable business (profit) models.

As we will see, these business models have generated some unintended consequences that fundamentally affect the value of the service provided.

Academic Search the Google Way

Ask a question. Set your price. Get your answer.

Search engines are having a powerful impact on libraries. Though unintended, this makes sense in that libraries are the traditional source for much of the information the search engines can now provide. The potential impact of search engine technology on libraries cannot be overestimated or ignored. The libraries know it. They are trying to determine whether the appropriate reaction is fight or flight. In 2002 Google launched a for-pay reference service called *Google Answers*. In response, the "free" reference desk department of a major Eastern university challenged Google Answers to a battle of accuracy. Though Google Answers was still in beta test (it has since emerged and is now available as a regular service) the challenged was accepted. What took place was a "junk yard wars" of information gathering. The results were fascinating and the implications for libraries chilling.

Auctioning Answers

The idea behind Google Answers is itself a fine example of the kind counterintuitive thinking that launches new businesses. What if, the Googlites mused, people could get students and researchers to *pay* for answers to reference-desk-type questions and, in turn, get qualified researchers to bid on the opportunity to provide the answers? A whole new information market could be created, a highly efficient exchange mechanism wherein the best and most knowledgeable reference desk "librarians" in the world could be available to anybody willing to pay. I put the word "librarians" in quotes because those providing answers would not have to actually be trained reference librarians. They could be anybody who knew the answer or knew how to find the answer to the question being posed. The best (smartest? most experienced?) people would work quickly and efficiently and be willing to answer for less money but still make more in the process. Researchers would benefit from better answers but would also be willing to pay for information because the efficiency of the market would keep response times short and prices low. Both parties to the transaction would benefit.

Here is how it would work. The customer would pose a question online. A number of reference "librarians" (possibly a large number, inasmuch as Google started with 800 pre-qualified researchers) would look at it. Some would decide that they could not efficiently or accurately answer the question. Others would be willing to accept the question and do the research. Since the customer would have already offered a fee for the answer (Google Answers' recommended starting price was $5.00 but any price could be offered,) the librarian would know approximately how much he/she would be paid (Google would take a percentage of the $5.00, of course, and also charge the person asking the question an additional $.50 just for processing the question through the web site.) If no librarian was willing to tackle the question for $5.00, the customer could raise the offered price or do the job him/herself. Eventually, Google Answers theorized, the market would establish a price range for questions of varying difficulty and on various specific topics. Customers could accurately estimate what their costs would be, and the reference "librarians" could develop specialties to help them answer faster and earn more money.

In fairness, it must be said that Google Answers was not the first to use real people to answer online questions. A service called "Ask Jeeves" launched a search engine that was supposed to answer almost any question typed into its web page, but, as you can probably guess, they immediately ran into problems with automated responses to many of the questions posed and so hired some people to answer those questions that confused the software. Initially, Jeeves implied that the software was doing all of the work, but it was soon discovered that people were providing answers in a sort of delayed live chat mode. The Ask Jeeves program was not successful but today Ask Jeeves survives as a shopping search engine that finds prices and links readers to e-commerce sites that sell whatever the customer is searching for.

Ask Jeeves saw great potential in online information services but their business model was anything but innovative. They offered free answers but hoped to sell advertising on their site. On the other hand, Google Answers wants to sell the information itself instead of indirectly supporting the information service though ad revenues. The challenge to Google Answers is not so much from other software but rather from the notion that reference desk assistance is available FREE from thousands of libraries around the world and has been since the advent of the written word! Why would somebody PAY for something when they can get it for FREE? Perhaps libraries have nothing to worry about.

Let's consider bottled water. Water is free almost anywhere in the world. At a restaurant, ice water is usually the first item served and there is never a charge on the bill unless the customer requests special tonic or mineral water from a spring-fed pond or pool in some remote land. All kinds of public places, from retail stores to sports arenas, provide free water fountains. While some fast food places charge a nominal fee for a cup of ice with water ($.15 is the most I have ever been charged,) the water itself is free. The fee is for the ice and the cup, and is usually waived if you buy a burger. Tap water runs in unlimited quantities in almost every house in the nation. Indeed, running water is a basic indicator of modern civilization throughout the world. And yet, the major soft drink companies have developed an enormous bottled water industry. It generates hundreds of millions of dollars in sales each year, all for a product that is readily available for FREE.

We could mention many other examples of free things that can be bought. You can even buy oxygen at specially equipped "bars." Attempting to sell something that is available for free no longer even qualifies as an innovative idea. Google Answers thinks that if somebody will pay $1.79 for a pint of Dasani (which is nothing more than filtered tap water) they will pay $5.50 (total) for an academic search or some other look-up conducted by a research expert. Google Answers also thinks that the expert will actually do the work for about $3.00 (what is left after Google takes its fee) and that the bid mechanism will keep prices low which in turn will keep the customers coming back for more. All of this sounds great, assuming of course that the answers are correct.

Battle of the Giants: Libraries vs. Google

This brings us back to the challenge made by the university reference desk. Both sides were professional and polite. The university described the study as a "modest" effort to "compare and contrast" the two types of services so as to learn why Google's approach and online search in general has become so popular. Google was gracious in saying that it would take "hundreds" of years before search engines would be as good as reference librarians. And so the battle was joined.

Three groups of test questions were used, including one borrowed from a different but similar study evaluating search engines and one composed by the university's reference librarians based on typical questions answered by them over the prior six months. The questions were emailed to both Google Answers researchers (which the university variously referred to as "freelance researchers" or "hobbyists") and to the university's reference staff. Other university reference staff then conducted a blind review of the responses. The results, which the university described as "more exploratory than scientific in nature" were statistically tabulated in five detailed charts. The university concluded, "...there was no clear winner."

The fact is that the university did not win nor did it even earn a tie. To their credit, they asked some hard evaluative questions including overall accuracy (value to the consumer,) consistency of response, quality and cost per response. Regarding accuracy, the university did better on the questions they composed, but Google did much better on the questions from the third party that was not affiliated either with the university or Google. Regarding consistency, Google's answers were rated as far more consistent that those of the reference librarians. Regarding costs, the university costs per answer were anywhere from two to five times greater than Google's. In the university's favor was the fact that two of the questions were considered too difficult for the Google's "hobbyists" to answer at normal rates and nobody bid on them. When the price offered was increased significantly (from five to twenty times the original price) Google's researchers answered the questions but no longer enjoyed a cost advantage over the university librarians.

It should be noted that there was no effort to measure actual "satisfaction" with the answers provided by either source. Accuracy and satisfaction are not quite the same thing, since all information provided might be accurate but the overall answer might be incomplete. In such a case the customer probably would not be satisfied with the results. Google Answers offers a guarantee with its answers; if a customer is not satisfied, all money (except the $.50 question fee) is refunded, no questions asked (no pun intended.)

Even though the university maintained that nobody won, it may see the writing on the wall. It concluded that, among other things, the results of the study suggest that at least some of the reference desk work currently being handled could probably be done just as well but more cost effectively if it was "outsourced." This would allow librarians to focus on more "complex information discovery functions."

What really should bother the university reference librarians is the fact that they lost badly when it comes to the *consistency* of their answers. Some explanations were given as to why this was the case, especially the fact that the Google Answers' researchers relied heavily on Google (and other) web search technology while the university reference *librarians were able to consult a much wider array of sources* which included printed materials at their library as well as online resources. Having more sources ought to be a plus for the university librarians because it gives them an advantage over web-only resources. But if the better resources are not available to ALL of the university's reference librarians, or ALL of the librarians don't know how to use these resources, then the advantage is lost and the quality of the response received by a university reference desk "customer" depends not just on the knowledge of the librarian on duty at that particular time (which can vary) but also on the specific resources available to that librarian at that specific location. This is a problem of analog resources (printed copy references) that could be solved by making all resources available in digital format so that they can be shared seamlessly throughout the university's library system. This would not only make answers more consistent (assuming consistent skills on the part of the various reference librarians, which can be achieved through continuous training,) it would also improve the response time in answering questions because of the faster speed of consulting online resources as compared with handling actual books. Of course, all librarians would have to be trained to use the broader array of digital sources with equal skill.

Planning a Rematch: Can Libraries Win?

The university librarians recognize the problem with consistency and advocate greater effort to improve research skills. They start with a call for greater "self-assessment" citing a study published in late 2002 that only 3% of the 77 responding libraries to an Association of Research Libraries survey regularly assess the quality of transactions. The most common statistic gathered was the number of reference desk transactions, which has shown a sharp decline in activity over the past couple of years. Ninety-nine percent (99%) of transaction statistics gathered were recorded with paper and pencil, creating analog data that is difficult to share and compare.

Search engines are also aware of their own limitations, the primary one being that they can only find information that is accessible through web sites. Companies such as Google and Yahoo are expanding search resources to include images, music, video, etc., and exploring the uses of Peer-to-Peer (P2P) networks as an aid in searching for information contained not just on websites but also in the hard drives of PCs connected to the global online network. This could exponentially expand information accessible to online search. Google has also introduced a search algorithm that focuses in on citations contained in academic papers as an aid to academic research. Google claims that this will help separate scholarly content from commercial web sites, thereby assisting students in their work. Google is offering this service free and without advertising as a way of "giving back" to the scholarly community.

Despite the fact that the library did not beat Google Answers, it seems to me that a winning strategy is at hand. Outsourcing may appear to make economic sense due to lower costs, but is not the answer. Researchers need reference librarians. What the library needs to do is offer *faster speed* (by putting all of its superior research resources in a digital format instantly available to all of its librarians,) *consistent, accurate answers* (through continuous personnel assessment and training,) all at *no cost* to the customer (reference is traditionally a free service at libraries from the community to university level, paid for by tax dollars or include in the cost of tuition.) The librarians have to be trained to use these advantages to their fullest. If libraries are faster, more accurate and still free, Google Answers does not look so formidable.

In summary, there has been a huge upsurge in the popularity of online search over the past couple of years, which has impacted the use of reference desks at libraries. In preliminary head-to-head comparisons with traditional library reference desk services, search engines have performed quite well. As more and more students use online search engines, use of library reference desk resources might continue to decline. However, libraries still have many advantages that, if optimized, could make them the better information source than online search.

Yahoo as a Research Tool

In order to gain a sense of how good Internet search engines really are (as opposed to a traditional reference librarian, for example) let's test a typical search engine to see how it performs.

Search is conducted when the searcher enters a word or phrase into the search bar and the software then matches that word or phrase with keywords (words and phrases found in web sites.) The search engine companies caution that results are dependant upon the degree to which the "structure and syntax" of the search phrases being used is optimized, so that if you learn a list of "search rules" and compose your query accordingly, you can improve the odds of actually getting the information that is relevant to your needs. In fact, tutorials on search engine strategies have become an online cottage industry, with experts giving advice on how to manipulate queries and coax the search engines into giving more exact or useful information. Google actually publishes a book (hard copy) on how best to use their search engine to achieve the desired results.

Let's run a test to see how this works. (Note: the number of "hits" — a hit being defined as a web page listing that contains the word or phrase being searched — from a search is never the same two days in a row, so that you will probably not get my exact numbers if you try this yourself.)

1. When I entered the word *mattress* in the Yahoo search bar, 3,600,000 "hits" were found. (At the default setting of 10 site hits per page, I would have to scroll through 360,000 pages of listings to view all listings!)
2. Switching to a narrower search query, *mattress pads*, I received 350,000 hits, some of which contain the word *mattress*, some of which contain the word *pads* and some of which contain both.
3. If I refine the syntax by adding quotations around the two words ("*mattress pads*") the search is narrowed to 214,000 hits because now the web sites listed must contain the phrase *mattress pads*. So even though the searches are identical except for the quotation marks, the total number of hits decreases by 38% when I search with the quotations.
4. Narrowing the search is also possible by adding words to further refine the subject. When I queried "*hypoallergenic mattress pads*" in quotation marks, I received 312 hits.

How the search syntax is structured definitely makes a difference and this language is something that has to be learned. It is not intuitive.

The Tendency of Keyword Search to Deliver Commercial Information

It is important to understand that the search engines can only see information that is contained in active, online web sites. The means that a great deal of information that might be available on a particular topic is not available to search engines. What the search engines find tends to be biased toward commercial information and toward web sites that provides online purchase options. This is inherent in the keyword selection process, which is manipulated by commercial web sites to assure that their web sites are listed on the first few pages of hits. Search engines encourage this as part of their business model, which caters to shoppers with the promise of vast quantities of valuable consumer information. The search engine sites also display paid advertisements along side of or even within the hits list. (I discuss this is more detail below.)

Although the Internet has grown in popularity as a source of consumer information, this is not always what people want. Even a query like "*hypoallergenic mattress pads*" might be made by someone merely looking for medical studies or other information about whether or not these pads have proven to be effective for certain types of allergies. Yahoo might find web sites containing this non-commercial type of information or not, depending upon the exact words used in the query. The flaw in using this method of searching for information would be that *you never know what it is that you do not see listed.* The best source might be buried hundreds of hits down the list, it might be invisible to the search engine (online, but password protected inside a database, for example) or not there at all (lacking the keywords to match your query.)

By the way, it is it difficult to compare the performance of the Microsoft search engine (MSN Search) with Yahoo or Google because MSN Search does not list the total number of hits found from a search; it merely asks you to click to see the next 10 hits.

Not Knowing Where to Start

What we have learned is that narrowing the focus of the search phrase improves the quality of the results delivered by the search engine. But what if you don't know enough about what you are looking for to create a narrowly focused search phrase? This is the same problem faced by grammar school students when they are just learning to spell and are being taught to look up the exact spelling of a new word in the dictionary. Let's say the word is "xenon." If you do not know how to spell it to begin with, you may be unable to even get started. You could try phonetically with "zenon" but there is no entry under zenon. Unless you can find the word spelled correctly somewhere or you know somebody who already knows enough about the word to tell you that it begins with an "x," your dictionary look-up will not work.

Using a search engine can be just as difficult as trying to look up a word that you cannot spell. How do you look for information about a topic that you do not know exists? Like the dictionary, search engines depend on the searcher having a substantial amount of pre-existing knowledge, at least enough to get the search started in the right direction. This is not always the case.

Does better software and "artificial intelligence" (AI) offer hope? Possibly. When I entered "zenon" in the Aspell Spell Helper the very first item recommended was "xenon." The search engine companies think AI or semi-intuitive software like

Aspell are the future of queries, although nobody knows when this will be available. In the meantime, there is no Aspell for the search engines. Except for the library "help desk" we have nowhere to turn.

User frustration and even outright anger with search engines has spawned some interesting, if unanticipated, reactions, including the invention of a new game called "googlewhacking" (The Search for the One.) It is a tongue-in-cheek effort to get cyber revenge for all the time that is wasted sorting through thousands of search hits. It works like this; two completely unrelated words are entered into the search bar without quote marks. No formal names are allowed. The goal is to register exactly one (1) hit, no more, no less. You don't win anything but you can get your one-word hit listed on a web site along with your name, so there are bragging rights. If it sounds easy, try it. I searched the combination of *bicycle* and *oyster* and got 51,000 hits! *Petrol* and *clips* got 34,500; *aardvark* and *union* got only 105 hits, *dustbunny* and *amoeba* got 34 hits, and so on.

Of course the search engine ultimately wins by destroying your prize combination. The whackers (as they call themselves) have discovered that as soon as a winning combination is posted on the googlewhack web site, the search engine finds it as well as the original hit, so now there are two hits and the word combination becomes a loser. Others may also mention the two words, and the search engine hits increase even more. At one time *francophone* and *namesakes* was a winner, getting only one hit. The number is now over 50. Cyber revenge, and the fame that goes with it, is fleeting.

An off-Broadway show called "Dave Gorman's Googlewhacking Adventure" opened in late 2004. A book of the same name has also been published.

Book Search with Amazon's A9

In September of 2004 Amazon announced a new search web site designed to compete with Yahoo, Google and MSN as a "one-stop" search platform. Called A9, it adds graphics (thumbnails of web pages located) and other "premium" features such as search history, an online diary, bookmarks (similar to Microsoft's "favorites") and search lists. None of these things are real breakthroughs. In fact, most of the searching is done by other search engines linked to A9, so it functions a lot like many other metasearch "search the search engines" (MetaSpider.com, for example.) In fact, A9 seems to depend on the Google search engine and the Google Images search engine, with help from GuruNet and the Internet Movie Database. It is unlikely that Amazon will beat them if they ARE them.

But, one aspect of the A9 site is worth noting by libraries, and that is the *Search Inside the Book* feature (formerly "Look Inside the Book") that was developed by Amazon to help customers decide which books to purchase. Here's how it works. Amazon provides a search engine that finds the customer's requested keywords on the Amazon web site (it does not search the entire Internet, just the Amazon site.) It lists any books available from Amazon that contain those words. The customer can then select a specific book and see references to specific pages inside that book that contain the keywords.

It gets better. Much better. For "registered Amazon customers" a deeper level of information is available. Amazon takes the customer directly to the page that contains the keywords she/he selected and the customer can browse forward and backward two pages in either direction with whole pages displayed. In other words, the keywords being searched can be read in context with up to four pages of text from the book. This greatly expands the customer's ability to find relevant information.

Of course what Amazon wants to do is sell books, not just facilitate contextual searches. They provide a convenient shopping cart so the book can be ordered online.

But seeing keywords or phrases in the context of a few full pages of narrative or text makes this a potentially far more useful search engine. What is unclear is how many books can be searched this way? And what kinds of books? General interest? History? Fiction? Entertainment? According to Amazon, they have "millions of pages" scanned and online (the exact number is not given) and they say they intend to "widely expand" the number of books available. They have not commented yet on the types of books they intend to scan, but we can assume that, since their goal is to sell books, they will focus on newly published items and not on older materials. They offer a Publisher's Resource Guide to help merchandise new books by providing excerpts through Search inside the Book. Publishers are asked to provide a free copy of their latest books to Amazon, who then scans them and adds them to their search archives. Amazon makes it clear that the whole purpose of this is to "sell more books."

It seems to me that by including Search Inside the Book technology into the A9 search engine, Amazon has greatly expanded access to information over the Internet by researchers. Their purpose is to sell books but there may be unintended consequences in the form of non-commercial benefits. By exposing the inner content of books (and not limiting the information provided just the author, title and price) they make this service far more useful to researchers using their search engine. This is a major leap forward. If they can scan enough books into their system (it is not at all clear that they can with the speed limitations of current scanning technology, but this is improving all the time) they could impact not just library reference services but also the demand for the physical contents (books) of the library. Amazon customers might decide to purchase the books instead of checking them out of the library, especially if the cost of the books was lowered through digital formatting and distribution. There would be no paper, ink or binding materials and no shipping costs since it could be distributed over the Internet. Amazon might not want to exert downward pressure on the price of books, but their technology certainly makes this a possible consequence of what they are doing.

On the other hand, customers might decide to research the books using the A9 web site and then go to the library to check them out for free. Since most of the books listed by A9 are new releases, the pressure would be on the library to acquire these books and have them cataloged and on the shelf quickly (Amazon advises publishers that they need 3—5 weeks from the time they receive a copy to make a book available online.) The value of a library to its students might depend upon whether or not it can select, order, catalog and inventory books as fast as Amazon.

Update: Google Bytes Back

On Wednesday, October 6, 2004 Google responded to A9 by issuing a press release announcing that it is launching a new service to make books and other printed materials searchable online. Publishers can send their books to Google, where they will be scanned and added to Google's search index at no charge. Users will be able to view information about a book and browse select pages, though not the whole book. Like Amazon, the pages will also contain links to online booksellers. Google said the service would start with "a limited number of English-language books and will expand to other languages in the next few months."

Google chose the Frankfurt Book Fair, the world's largest, to make the announcement.

4. Search Engine Problems

The appeal of online search technology is undeniable as evidenced by the fact that over 60 *million* Americans per day were using the service by the end of 2004. However, there are many problems with the way search engines gather and display information and also with the information content itself. Chief among these is the pervasive and still growing influence of advertising, which I discuss in Chapter 5, below. But there are many others as well. There are also huge gaps in the information search engines are able to provide. Often these problems and gaps are not readily detectable just by testing or using the search engine itself. This makes them harder to recognize and understand and in many ways more dangerous because users, especially younger, inexperienced or untutored searchers, may be unaware of the potential limiting or adverse effects. Here are some of the problem areas that are most challenging.

1. Offline Content: Inaccessible Analog Information

Search engines cannot see information that is non-digital. Printed books are the best example, although images, video and music are also un-searchable. There are billions if not trillions of pages of books that cannot be accessed over the Internet. All of this information is of interest to the public and available somewhere, somehow (usually libraries.) But it is invisible to Internet searchers. Search engines are hoping to bring increasing amounts of the non-searchable information within reach over the next few years. Google has launched a program to scan (convert to digital format) library content for seven major U.S. libraries, including the libraries at the University of Michigan, Stanford, Oxford in England and Harvard, and the New York Public Library. The project is expected to take many years, but the first work was made available online in mid-December, 2004. Harvard has agreed to only a limited number of books being scanned as a test (40,000 of their approximately 15 million volumes) but Oxford wants all pre-20th century items scanned. According to a Google press release, "the goal of the project is to unlock the wealth of information that is offline and bring it online." According to Paul LeClerc, president of the New York Public Library, "It could solve an old problem: If people can't get to us [libraries,] how can we get to them?" A University of Michigan librarian spokesperson added "It will be disruptive because some people will worry that this is the beginning of the end of libraries. But this is something we have to do to revitalize the profession and make it more meaningful." Michigan estimates that it will take six years to scan the 7 million volumes in its library. Cost estimates for this project are between $10 and $20 per book and Google has agreed to pay for the entire project. Stanford wants a "substantial number" of its 8 million volumes scanned as part of the program. In an important departure from their usual business model, Google will not have advertising content or links accompany the search results, at least not right away. Instead they will link to Amazon.com where books can be purchased, and they also plan to *link to libraries where the books can be borrowed for free.* The say this will build a cooperative relationship between analog sources and Internet search engines. Libraries hope to increase their traffic but putting their content online, and retail books stores hope to gain customers even though their exposure comes in an ad-free environment. It is a good example of counter-intuitive thinking that makes the Internet so promising as a serious (as opposed to purely commercial) information resource.

2. Offline Content: Inaccessible Information in Local Systems

Microsoft, Yahoo and Google and other search engine developers think the next breakthrough in search will be local hard drive search (called desktop search,) not Internet search. Users now save so much information in their individual PC's (in the form of emails, documents and instant message files as well as stored web page links) that they often cannot even find what they already have accumulated. Regular Internet search engines cannot access this information. Yahoo, AOL, Ask Jeeves and Google have announced plans to introduce (or have already introduced) PC search software that looks for keywords designated by the user, similar to Internet search engines. Microsoft's program, called MSN Toolbar Suite (currently in Beta) allows keyword search within .pdf files as well as text and html files and spreadsheets and can also search network hard drives, which the others cannot.

Desktop search is already controversial due to privacy issues. Google's program, for example, sends your search query to Google and, while the software is scouring your PC for the requested files, Google analyses your query and sends back related advertising links that are displayed along side your requested files on the search page. This means that Google monitors every search you conduct for any advertising advantage it might create. Theoretically, Google could accumulate information about you through your queries and then send other types of ads, pop-ups, for example, that were appropriate to your business activities or lifestyle.

All of the desktop search programs have similar issues, although all deny that the information they gather will be misused. Also, none of this software would enable Internet search engines to access local PC's. Desktop search is still in its developmental stage but it is expected to be a finished product by late 2005 or 2006.

3. Offline Content: Inaccessible Digital Information

Search engines cannot find information unless it is available via a web page that can be accessed via the Internet. There are untold billions or even trillions of pages of information, such as searchable databases, technical and scientific reports, business and academic studies, notes, letters, diaries, government documents, etc. These documents may already be in digital format (plain text files, databases, Word documents, Excel spreadsheets, Power Point presentations, emails, etc.,) but they are protected or hidden behind a firewall or otherwise secured from public view inside of corporate or government Intranets, for example. Many are User ID or password protected, so the automated "web crawlers" that find available web sites and add them to the indexes of search engines, cannot enter the sites. Many searchable databases, when searched, create dynamically generated pages that respond to a specific search query. Search engines can only find keywords in static pages that are stored on a server, so that even if they can look into the databases, they are unable to search database content in depth. Much of this information may be legitimately proprietary or subject to security or privacy considerations, as determined by businesses, governments and individuals. But some searchable databases store information dynamically not to keep users away but merely to save space. The decision as to whether information should or should not be more readily available to researchers using the Internet rests with the owner, company, agency, copyright holder, etc. But access to even a small part of this information would increase exponentially the amount of information available to researchers, so the Internet search companies are looking at access options.

One possible solution that is being explored by search engine companies would allow the search engines to "see" information voluntarily placed into peer-to-peer P2P folders by the owners of the information [for more on P2P, see Chapter 6 on "Ambient Information."] This would leave decisions regarding accessibility up to the information owners or custodians, thereby avoiding conflicts over invasion of privacy. But if even a fraction of these private documents were voluntarily placed in P2P folders, the Internet would be enriched significantly, and without the advertising content that characterizes most online web sites.

4. Online Content: Reliance on Keywords

There is ample evidence that the customer-directed nature of digital search engines limits their effectiveness. Search engines are heavily dependant upon the quantity and array of keywords imbedded in each different web site available (there are now over 8 *billion* web sites online and the number is increasing daily.) This is because the search engines return site selection to the searcher by using their web crawlers to build word and phrase indexes and then cross-reference key search words provided by the searcher with keywords and phrases from their indexes. Search engines also use other criteria for search selection including frequency of visits by users and links to sites from other sites, but keywords are the primary search mechanism. Web designers may or may not do a good job of imbedding keywords, and many sites have no keywords at all. This forces the search engine web crawlers to look deeper into individual site content (if they can gain access.) This is one reason why a simple customer query, the word "mattress" that we used in the Yahoo test for example, yields over three millions hits, sites that may or may not contain useful content but at least contain the word "mattress."

Linguist Geoffrey Nunberg points out another serious problem with search by key word, that being the inadvertent filtering of information due to preexisting bias. For example, a Google search using the query "abortion" would miss many web sites that consciously avoid use of that terminology and instead use the term "pro choice." Careful word selection by advocacy groups is a critical to their message and they avoid use of many terms that they deem derogatory. Pro life advocates never refer to themselves as "anti choice." This affects their key word list. The searcher may have an open or subconscious bias or may unknowingly select a search phrase that is favored or avoided by one side or the other of a controversial issue. Whatever the reason, the phrase selection filters out hits that might otherwise be valuable to the searcher. Nunberg thinks that a researcher exploring the same topic in a library would be more likely to encounter diverse points of view.

Keyword-based search has another problem that being the use of Spamming techniques by commercial web sites to get themselves placed on hit lists. By embedding all of the popular (generic) keywords used in search (these are carefully tracked and reported by the search engine industry) or other keywords that are not appropriate for their site, commercial sites can become part of completely unrelated searches. There is no mechanism or incentive for search engines to deal with this problem, which has the effect of further cluttering search results with irrelevant information.

The World Wide Web Consortium's Semantic Web project is trying to change web standards to improve the ability of search engines to find and understand the keywords a web site uses to describe itself. However, any significant search enhancements are several years away.

The alternative to keywords and phrases is to allow searchers to type complete senetences and longer, more complex phrases as part of the seach query. Called *natural-language queries*, this type of search goes beyond the massive indexes assembled by the search engines (these indexes catalog keywords and phrases to speed the search process.) Proponents of natural language queries claim that indexes limit the information responses available to searchers and that natural-language queries go deeper into the actual document content of items indexed, resulting in a better search response (meaning more useful information, not more hits.) Ask Jeeves pioneered natural language queries and says they will re-introduce technology sometime in 2005 that will further the question-and-answer abilities of their search engine. The new feature, *Direct Answers From Search*, will search across the entire web, rather than simply from the Ask Jeeves database, to find answers to natural-language queries (that is, those phrased as questions rather than mere keywords.) Whether this actually improves on keyword search remains to be seen. Critics of natural-language queries point out that hundreds of millions of searchers have already been trained to enter one or two keywords and will be unlikely to adapt to a more complex search methodology that requires them to write complete senetences.

5. Online Content: Identifying Useful Information

Yet another major problem, encountered frequently by search engine users, is that there is no way for the search engine to list the located sites on the basis of *usefulness* to the searcher. The mattress pad example used to test Yahoo is a good example. Was I looking to purchase a mattress pad or was I doing biomedical research on allergies? But there are usefulness issues (referred to as "relevancy" by advertisers) even if all I want to do is buy a mattress pad. Geography (referred to as "localism" by advertisers) is an example. Finding a web site that features a retail location in Pal Alto is useless to me if I live in Boston. The best site for a customer's specific need may be scores if not hundreds a pages into the list, making it impossible to find. The sheer volume of information, randomly sorted, makes Internet search engines far less effective than users would like them to be. Work is under way on this problem but, as with keyword search enhancements, results are preliminary.

Let's look at three interesting examples of solutions to the usefulness issue, each taking a somewhat different approach. They are all currently in beta (test) mode. Whether or not they catch on as an improvement over Yahoo, Google, et al remains to be seen.

Blinkx

One promising example is a new utility program called "blinkx" (www.blinkx.com) which continuously searches both the Internet and your local PC for web sites and stored files and emails that relate to words in your Word or browser document. The idea behind blinkx is to reduce the use of keyword-based search and instead use what is known as contextual search. Blinkx assesses all the information that you are actively viewing, and automatically recommends and retrieves relevant content, including PDFs, Zips, MP3s and JPEGS from local and Web searches, *based on context*. So if a student is viewing documents on Martha Washington, blinkx is able to extract the main ideas contained within these documents and understand what is being researched (in this case Martha Washington and early American history) using what they call Context Clustering Technology (CCT). Their stated goal is to overcome the shortcomings of keyword technologies that, in their view, lack context and produce a jumble of thousands of irrelevant hits during a search. They want the search engine to truly understand the ideas and context behind the words. Blinkx has several useful innovations including Smart Folders, which automatically search the Internet and update content as new information becomes available. They do this by knowing the context of the documents already in the Folder. No additional keyword search is required. A toolbar allows you to quickly check what blinkx has found, so that some semblance of continuous research occurs as you compose and edit a research report, term paper, etc. With a computer online 24/7, blinkx serves as an information bot (short for robot) that can search and gather hits whether or not you are actually using the computer at the time. Blinkx limits its results to about ten items per search. Blinkx has also teamed up with Movielink to add search capability for online movie trailers using voice recognition to create a searchable index of movie dialogue. Unfortunately, blinkx is only in the prototype stage and no one knows when or even if it will prove reliable enough to be launched commercially. Also, blinkx needs very high broadband speeds to keep pace. These are unavailable throughout much of the U.S. but that is expected to change over the next few years.

Snap

Another interesting approach to assisting a searcher in find the best sites for a particular query is to allow sorting of search results for an individual user and then combining that with some qualitative rankings based on feedback from prior users of that site. This is being tested by a search engine called "Snap," (www.snap.com) which was launched in beta test mode in January of 2004 and introduced commercially (it supports itself with ads and sponsored links) in late 2004. Like other search engines, Snap finds web pages based on an initial query supplied by the user but also ranks each web page found by several additional criteria such as popularity (the number of visitors from the Snap Network of Internet users who clicked on this specific web site, which Snap thinks will help the searcher make better selections) and user satisfaction (the average number of pages viewed by a visitor to this site, Snap's assumption being that the longer previous visitors stayed on the site and the more pages they viewed, the better the web site must be.) Snap also allows an additional query (filtering) of the web sites found by the initial search criteria to allow the searcher to bring more relevant web sites to the top of the list. This helps sort the hundreds or thousands of initial hits into more a useful list. Snap still depends upon keyword search and is limited to listing available on the Internet, however. Snap claims that 250,000 users tested their site during the first week of their beta launch. If this number is valid, it is indicative the intense interest among Internet users for more useful search engines.

Clusty

A different approach to solving the usefulness problem has been developed by a search engine called "Clusty" (www.clusty.com) which was launched in late 2004. Clusty does not search the web directly but rather searches a group of other search engines who are searching the web (Lycos, MSN, Yahoo, etc.—this is called metasearch technology.) Using artificial intelligence (AI) Clusty groups search results from these other sites into different categories. Like Snap and blinkx, Clusty tries to help the searcher sort through the huge number of hits that a typical search might make. This can be a real time-saver, especially if the search turns up a mixture of results based on different meanings of the same term. For example, if we enter the word *whitewater* we could be looking for information on kayaking, or whitewater rafting, or Whitewater University in Wisconsin or the political scandal involving former President Bill Clinton and his wife. Clusty sorts the hits into groups (clusters, which is where the name Clusty comes from) by using artificial intelligence to pick out the major themes found within the results for each search and then sorting them into folders. These folders appear on the left side of the computer desktop. The searcher can select the best one for his/her purposes and ignore the rest. For example, when we enter *whitewater*, one of the folders that is created is named "Scandal." If we are still battling the Clintons after all these years, this saves us a lot of time.

Clusty has another feature that is even better. There is a series of tabs across the top menu bar labeled *Web, News, Images, Encyclopedia, Blogs, Gossip, Shopping, eBay* and *Slashdot* (an online tech website whose motto is "News for Nerds.") These lead to additional search results from a variety of sources. Each of these tabbed sources is automatically sorted into folders as well. This is a tremendous help because it places a lot of the commercial material in an area that can either be accessed or ignored, depending upon whether the searcher is looking to buy something (a whitewater canoe, for example) or looking for other kinds of information (about Bill and Hillary.) When *whitewater* was searched, the Blog search created folders named "Investigation" and "Travelgate" (a related Clinton scandal) and clicking on the "Images" tab brought up about 6500 photos related to the whitewater theme (mostly recreational, none of the Clintons) sorted into convenient folders. Each photo, displayed as a thumbnail, has dimensions in pixels and total size in kilobytes and shows the URL of the website where the picture is located. This enables the searcher to look for additional information on the photo by going directly to the website.

The "Encyclopedia" tab is especially interesting because it searches the online *Wikipedia*, a public domain encyclopedia project that can serve as a useful starting point for academic research projects. The *whitewater* search turned up 209 Encyclopedia articles including 23 that were placed in a folder called "Scandal."

Clusty was developed by a company called Vivisimo, which was founded with a $1 million grant from the National Science Foundation, an example of our tax dollars at work. America Online (AOL) has licensed search technology from Vivisimo and the upgraded AOL Search Engine (released in early 2005) bears a resemblance to clusty but adds a new feature — sorting by geographical location.

6. Online Content: The Impermanence of Information

The information available over the Internet is often temporary, both in the sense that it changes and that it also disappears over time. Web and blog sites are often removed by their authors or are abandoned and the web site hosting companies delete them. News sites leave stories up for only a brief period of time and may or may not retain them in an archive. Newspaper archives are rarely complete. Advocacy, educational and other similar information sites often revise data or content and make no effort to save prior information or versions. A good example of this is medical information sites, where information that suddenly proves less than totally correct is "scrubbed" from the site and purged from the server. Servers themselves crash and data and even whole web sites are lost. Sometimes they are not rebuilt or replaced. Within web sites, links to deeper pages or to external references often break or point to information no longer available. This can be a problem with web sites that make the reader go to the home page (to see the advertising) and block deep links. In cataloged collections, changes to content require updates to indexes for search engines to work. This is often not done properly or lags content update due to workload. Search engines themselves do not always yield the same information, even on the same day. Most serious web searchers have experienced the frustration of seeing a link that looks interesting among a long list of similarly interesting links, making a mental note to come back to it, failing to write down or otherwise capture the Uniform Resource Locator (URL, the *www* address) and then never being able to find it again, even with the identical search query. Web sites sometimes change URLs and saved addresses are no longer valid. Many web sites that move may offer forwarding links, but some do not. It is unrealistic to expect researchers to print screen shots, download every web site or print out copies of every chart and spread sheet in order to guarantee that they can re-access important or interesting content. Printing out everything you find of interest on the Internet just so you can be sure you can find it again substantially undermines the value of search engines and of the Internet itself.

More traditional research sources have problems, but not these. Books may get lost or be replaced in the wrong location on a library shelf, but page 308 of a monograph will likely be there 50 years from now with the exact content it had when it was originally printed. Better yet, at least a few thousand copies of every book are printed, so that even if a specific copy is misplaced or destroyed, many more identical copies are available. (Web page backups are usually available only from the original owner. There are usually no duplicates available online as backups.) The redundancy and permanence of books means that information can be re-checked by authors and during peer review. Critics can more easily challenge the underpinnings of scholarship based on alleged misuse of data. Authors can more easily refute criticisms from pundits. Specific references can be made back to sources using standard footnotes or bibliographical annotations, knowing that the sources are still there for everybody else to see. By comparison, web-based information slowly melts before our very eyes, rendering search engine results less trustworthy and consistent than student and serious researchers would like. And what goes away does so without leaving a blank spot on a shelf, so researchers are unable to determine even what it is (or was) that is no longer there. They can only access what is presently available.

7. Online Content: The Accuracy of Information

The issue of accuracy and reliability comes into even sharper focus when we consider online collaborative web sites, such as the collectively written encyclopedias called *Wikipedia*, which are popular sources of information for high school and college students. Ordinary encyclopedia articles are peer-reviewed and held to certain standards for scholarship. Not Wikipedia. Although the authors collaborate and correct each other's work, there is no guarantee that anybody writing on any particular subject is qualified to do so except for the fact that they have an interest in the material. In fact, anybody can submit articles or modify the content of existing articles, regardless of qualifications or motive. There are some attempts at self-regulation but progress is slow. One particular collaborative site, *Wikinews*, lets anyone submit articles but they are posted only after a readers' review panel declares them to be "objective." There are already lawsuits involving both rejections and acceptances. Again, this is not something that concerns the search engines. They are being paid to lead the searcher to these sites and then their job is finished. It is up to the searcher to determine if the information is valid. This can require a great deal of time and work (actual research!) if done properly and negates the huge time savings and ease of finding information that makes search engines so popular to begin with. For this reason, there is great temptation to accept information provided as being accurate and scholarly, even if there is no corroborating evidence that it is. Even serious scholars can be seduced by research techniques that are quick and easy.

8. Online Content: The Issue of Searcher Privacy

Because search engine companies have access to all specific keystroke sequences entered by a searcher, the opportunity has been created for invasion of individual searcher privacy. Like everything else associated with search engines, this is being driven by revenue from consumer advertising. If a searcher keystrokes in an inquiry about hypoallergenic mattress pads or Datsun 240Z's, the search engine can instantly flood the page with ads for related products such as bedroom furniture, disco music or automobile memorabilia. Ad agencies call this "behavioral targeting." Though annoying to the searcher, it seems harmless until you consider the possible uses of the keystroke information that makes behavioral targeting possible. Suppose I enter queries about "cancer" or spend time on a job recruitment site. These keystrokes are also captured and they may trigger appropriate advertising, for oncologists, for example. But they may also be sold to life insurance companies or to my employer, who wants to know if I am looking for a position with another firm. I might be refused life insurance or not assigned to a desirable career-building project because of concerns about health or my commitment to the company. I would not even be told why, so that those using this information could do so without repercussion. Of course, my query about cancer could be related to my grandmother, and the job search could be for my daughter, something the keystroke tracking could not determine. But the damage to me could be severe and I would have no idea why. This is not merely a hypothetical problem; search engines and also web portals that invite readers to click on display ads, can do this now. Given the brutally competitive search for online ad revenue by search engine and news portals, there is no reason to believe it is not happening.

9. Online Content: Satisfied, Overconfident, Uninformed Searchers

In January 2005 the Pew *Internet & American Life* project issued a new study entitled "Search Engine Users," the latest in a series that began in 2001. Search engine companies should have been pleased with overall results because respondents report growing use of online search and a high degree of satisfaction with their search experience. Eighty-seven percent (87%) reported that they were successful in their search "most of the time" and 17% (included in the 87%) said they "always" find what they are looking for. This can only be described as outstanding customer satisfaction.

The functionality of the search is designed to assure a good customer experience. Search is fast and easy; it can be reduced to a single typed word in the search bar, which does not even have to be spelled correctly (the search page catches misspellings, automatically corrects them and diplomatically asks the searcher if the corrected spelling is what they really intended to type.) Unless the word is completely unintelligible, search always gets some sort of result, so that the user is always "successful." Google even offers an option (the "I'm Feeling Lucky" button) that selects the "best" web page of all those found (as defined by Google,) thereby helping to clean up the overwhelming clutter of thousands or even millions of hits that routinely occur with simple search queries (the fewer the words in the search, the less refined it is and the more hits occur.) The vast majority of searches are focused on a handful of topical areas. According to Pew, the topic five topics are (1) people, places and things, (2) commerce, travel, employment and the economy, (3) computers or Internet and technology terms, (4) health or sciences, and (5) education and humanities. This enables the search engines to concentrate their index building on a narrower range of web pages. This increases the number of available pages. Quantified by the *billions* of indexed pages available, at the end of 2004 Google reported 8.0, Microsoft 5.0, Yahoo 4.0 and Ask Jeeves 2.3. The larger the number of indexed pages, the greater the access speed, which has been reduced to a fraction of a second for most searches. Instant response further heightens the user's sense of accomplishment and satisfaction. Most searches are either one or two words. Although they vary somewhat from search engine to search engine, the top ten queries directed to Ask Jeeves during the first week of October, 2004 were online dictionary, music lyrics, games, Halloween costumes, jokes, baby names, quotes, Britney Spears, Paris Hilton and poems. Other sites reported heavy activity on airline flight tracking, KaZaa, Register to Vote, Mt. St. Helens, Clay Aiken, Pamela Anderson, NFL, Drudge Report, beheadings in Iraq, etc. (Pornography search statistics are excluded.) Search engines report the most popular searches on their web sites. Yahoo has what they call a "Buzz Log" that tracks the ebb and flow of searches. It is updated daily. Google's equivalent is called "Zeitgeist."

The same Pew report highlights what they called the naïveté of web searchers when it comes to hits and content. Search is the second most popular online activity, behind email, and the vast majority of searchers think they know what they are doing. Of 108 *million* Americans who have used search engines, 92% say they are "confidant" of their searching abilities with 52% saying they are "very confidant."

But when it comes to understanding what they are looking at, the Pew study uncovered some serious problems. For example, when asked about identifying the influence of advertising on search results, 68% describe search engines as "fair and unbiased" as a source of information with 19% expressing any concerns about this. But only 38% of those surveyed were able to identify advertising on the search site and only 18% said they could "always" tell when an advertiser paid for a hit. While 70% said they could accept paid advertising if it was clearly identified as such, 45% said they would stop using search engines if ad presence was disguised. As we will see shortly, advertising influence is heavily disguised. Searchers just don't recognize it.

There were some interesting differences in results between those under 30 and the surveyed group as a whole. Whereas 92% of all those surveyed expressed confidence in their search abilities, that number rose to 97% for those under 30. Twenty-seven percent (27%) of those under 30 reported using search engines "several times a day," 72% under 30 said that search engines were fair and unbiased (vs. 68% of the whole survey group) and 36% of the under-30 group said they "couldn't live without" search engines. In all respects, the younger users were more trusting of, dependent upon and enthusiastic about search engines, but they were not any more knowledgeable than older searchers about the content or functionality of search engines.

Search industry proponents and advocates such as www.imediaconnection.com reacted sharply to the Pew study, indicating that it had created a "stench" and was unfair to the industry. Understandably they were really nervous about the 45% who said that, if they detected advertising in a search engine's hits, they would stop using it. Industry supporters blame poor spelling of search queries, quoting pre-Internet comedian Will Rogers — "Nothing you can't spell will ever work." I am not making this up. I am not even saying they are wrong, although the idea that unsolicited commercial search results are caused by misspelled queries seems far-fetched. However, a Nielsen Norman Group survey (a much smaller sampling, it should be noted) released at about the same time as the Pew survey attributed the poor search results achieved by teenagers to "poor reading skills, unsophisticated search strategies and dramatically lower patience level." Maybe Will Rogers was on to something.

Spelling aside, the overall problem is not the fact that advertising is part of the Internet but rather that younger search engine users are less aware of advertising or advocacy bias and more trusting of information from search engines and the Internet. They are also much more likely than older users to use online resources for research, and they do so much more frequently. An awareness of content bias should be a significant factor in the use information for research. Apparently, however, it is not.

5. The Role of Advertising in the Search Engine Business Model

One of the most striking changes to the Internet over the past several years is the extent to which it has become saturated with advertising. Almost all web sites, including search sites, are commercially sponsored in some way or another. This is a lot different from early Internet days when advertising messages were banned (USENET, as an example). And as recently as 1998 advertisers could not buy paid listings that would appear along side of unpaid (organic) search results. But customs have changed. Search (and most everything else on the Internet) is now commercially driven. Even those pushing the web toward increased commercialism are a bit wistful about the old days. A Google executive, commenting on the upsurge in complaints about advertising in search results noted that "Even three years ago (that would have been in 2002) the Web had much more of a grass-roots feeling to it."

The influence of advertising by commercial companies on search engines *per se* is a result of the business (profit) model search engines have developed. Search engines do not sell products. They provide information services, more like a news service and, like news providers, they-are expected to provide their services free-of-charge to millions of customers each day. This is an unsustainable business model. So, like news providers, the search engines pay the bills by selling advertising. Google, for example, generates 100% of its revenue, currently between $4 and 5 *billion* per year, from ad revenues. In fact, the search engine business model, thanks to the successful IPO (Initial Public Offering) by Google in late 2004 and the extraordinarily high ad profits reported by Yahoo for the 4th quarter of 2004, has become a leading Internet profit center and a darling of Wall Street.

Money for Nothing and the Clicks Are Free

A close look at their business model reveals how simple it is. They make money by giving away information to customers and then charging advertisers to deliver ad messages to those same customers. The information they give away consists of free listings in the form of hyperlinks to millions of web pages. They can afford to do this because they do not pay for these listings. The web links are themselves free and readily accessible over the Internet, so they are giving away for free something that is already free. (In accounting terms, their cost-of-goods is zero, which is as good as it gets.)

Of course nothing is completely free. In order to enable searchers to find relevant links quickly and easily, the search engines have spent hundreds of millions of dollars and more developing sophisticated, software-based search technology. But none of this cost is passed on to the users, who enjoy almost instantaneous search capability for free. Instead, the search engines recover their costs through ad fees. What determines the ad rates paid by a company? The number of visitors to the search engine site (called a portal) who then "visit" that company's web site by clicking on a link provided by the search engine. The more clicks, the more advertising revenue. This puts a premium on efforts by the search engines to "build traffic," that is, increase the number of people visiting their site. In turn, the search sites try to pass these people on to the commercial sites listed on their search page. That way, everybody builds traffic. A certain percentage of these click-through visitors will buy something. It is what in sales is called a "numbers game."

On the surface, none of this seems to be a problem. Most forms of media—TV, magazines, newspapers, etc, do the same thing. They make their product useful or interesting to lure viewers or readers, and then count the number, and base ad rates on the count. Everybody has heard of the Nielsen Ratings and Sweeps Week. The appeal of advertising through a search engine is that the advertiser pays for actual results (click-throughs) instead of just having ad placements appear on the page. By comparison print media or TV ads have to be paid for even if nobody responds. Search engine click rates can be high, especially for generic terms. A rate survey by iProspect.com for the third week of January, 2005, showed that the keyword phrase *personal bankruptcy* cost $1.15 per click through, *background check* cost $1.19, and the term *refinance* was priced at $4.46. All this money changing hands has created the new crime of "click fraud" which occurs when bogus click throughs artificially increases the fees advertisers are charged. But the search engines claim that they deliver real bang for the buck (greater efficiency for a company's advertising dollar, in economic terms) for these astronomical rates. Due in part to this extreme efficiency, which advertisers like, the expansion of this kind of advertising in coming years seems inevitable.

And what do searchers find when they click on a paid advertising link? Not surprisingly, they find commercial web content, which is probably what they want or they would not have clicked on an ad link in the first place. (The web address can also provide a clue-"com" in dot-com stands for "commercial"—although .net and even .org have heavy commercial content.) A great many web sites, e-commerce sites for example, are obviously commercial. Others provide information on products and services that can be purchased at retail outlets. These sites may also deliver "pop-up" ads to accompany the web page. (Pop-ups have to be closed separately and therefore linger longer in front of the reader, which hopefully generates greater interest.) The key here is that searchers who knowingly click on an ad site should expect to get advertising with whatever information bias that goes with it. The goal is to sell products to people who have expressed interest in buying those products by clicking on a sponsored link. Advertising and advocacy are, by definition, unlikely to provide fair and balanced, peer-reviewed, comprehensive information. They are only interested in selling things.

The Influence of Unidentified Advertising on Search Results

As search engines have worked to perfect their business model, new kinds of advertising strategies, much less obvious, have become a part of web search. In fact, some are virtually invisible (pun intended.) For example, as we saw in the section on "Yahoo As a Research Tool" in Chapter 3, advertisers now pay fees to search engine companies to influence ad placement in a couple of different ways. The newest technique is what is called *relevance*—the relationship of the ad to the specific user query. If you search "Yogi Berra" you are likely to see placement ads for many kinds of sports memorabilia, not just Yogi. You may also see ads for books on sports, for authorized team sportswear, baseball fantasy camps, etc. Unfortunately your monitor may also be cluttered up with ads for children's pajamas imprinted with the "Yogi Bear" cartoon character. Obviously, this is not relevant or even closely related to the search. Due to the speed required to make relevant ad placements, the process is totally automated and is not an exact science. Nielsen surveys show that the accuracy of relevant ads has actually declined from about 80% in 2000 to 56% in 2003. This is significant problem given the limited amount of as space on a search portal. Search engines don't dispute the result but argue that relevancy and the measurement of relevancy are "subjective." But the effort to make advertising relevant to the search being conducted, though imperfect, has become a significant influence on the paid ad content displayed.

Another strategy for search advertising is based not on the actual sponsored (and therefore clearly identified) placement ads on the site but rather on the *sequence* (called rank) in which hits (the results of the searcher's query) appear on the search page. This takes advantage of the tendency of searchers, who may get hundreds of thousands or even millions of hits on a query displayed 10 to the page, to click on the first few on the list. Sponsored ads, which are more readily recognizable as ads and are located separate from the hits list, can be skipped. But unidentified ads surreptitiously placed at or near the top of the list of search engine responses to a query may get clicked just because the searcher assumes that results are from the query. The major search engines such as Yahoo, Google or MSN actually have a vested interest in not sorting listed web links into more useful categories so that users will continue to click on the first links they see because, as one student told me, "you've got to start somewhere." (I discussed some of the efforts by lesser-known search engines to sort data in the section of Chapter 4 entitled "Online Content: Identifying Useful Information.") And so the ads that pay to appear at or near the top of the search hits list are not identified as ads. The searcher thinks they are un-sponsored (organic) search results and clicks. Once again it is a numbers game; a certain percentage of click-throughs will make a purchase.

Manipulating rank to place unidentified ads in front of searchers has created yet another ad opportunity for search engines, based on a modified version of the theory of relevancy. These are links (paid for but not identified as advertising) to products thought to be of interest to the customer even though the words or phrases in search query would not ordinarily indicate that the customer might be interested in that information. These sponsored links appear right along with the normal site hits that result from search, not in a separate area of the web page, and so they are hard to identify as ads. For example, lets say I am interested in 1970's vintage sports cars and do a search on "Datsun 240-Z" which was a cool two-seater back in the Age of Disco (Datsun was the brand name formerly used by Nissan.) The search would no doubt turn up many appropriate sites but I might also see a link to a web site that sells '70's music, the idea being that the nostalgia for the Z-car might also trigger a yearning for the Bee-Gees. Search engines accomplish this by playing off the keywords I enter into the query bar on the search engine site. Typing in the word "Datsun" leads to "cars" (which it should), which leads to "1970's" (which is somewhat relevant), which leads to "Bee Gees" (which is irrelevant.) But the fact is, even though I was not thinking about music at the time, I might spontaneously click on the link and take a quick look at the Bee Gee's discography. This is just another version of the numbers game; the advertiser knows that a certain number of people who click-through will buy a CD. Is this beneficial to me as an individual searcher? Will it start a disco renaissance at my house? Or will this merely create a clutter factor that makes my search less efficient? At the very least this would seem counterproductive to the goal of the serious researcher, who would probably be more interested in the speed and accuracy with which requested information was delivered than the availability of classic disco music. A search engine's unwillingness to focus on what is relevant may actually drive users away from search engine sites. (This is an unintended consequence. As such, it creates an opening for competitors.)

There are also other advertising-related strategies being used by search engines that are troubling, including a tentative foray into keystroke tracking which has raised issues of confidentiality of searcher information. (I touched on this issue in the section of Chapter 4 entitled "Online Content: the Issue of Searcher Privacy.") But the listing of web pages found based on who is paying for the top rankings on page one is a growing influence on search results for millions of searches each day. The problem is that there is no good way for the searcher to tell that this is happening unless someone shows them what to look for.

Advertising has also begun to penetrate the "blogosphere," one of the Internet's newest information sources, a place not thought of as a haven for ad messages. Blogs are essentially web pages and are found by search engines just like any web page. They began as little more than an oddball form of personal psycho-cyber-diary with a small readership but came into their own during the U.S. elections of 2004, when their legitimacy grew dramatically. As political blogs became popular, millions of new readers logged in to get information updates and opinions on the various campaign issues. The reputation for accuracy and reliability of political blogs was enhanced by a string of news scoops throughout 2004 combined with comments and analysis by thousands of "regular citizens" (dubbed "pajama-hedeen," signifying their irregular status as reporters but also their intense commitment to issues.) Blogs have been credited with injecting some real grass-roots democracy into the democratic process, and they gained even greater exposure and credibility in the aftermath of the tsunami that devastated the coastlines of several East Asian countries in early 2005. It is estimated that 50 *million* Internet users now read blogs regularly (in 2002, the estimate was not more than 100,000.) About 12,000 new blogs are created *daily*; the estimated total number of active blogs is now several million. Blogs have gained the reputation of being a medium for independent exchange of information between or among voters, consumers, on-site observers, political pundits, etc., enabling readers to publicly praise or criticize candidates or commercial products and services based on their own, everyday experience. Many politicians and corporations have been raked over the coals in public though blog sites set up specifically for that purpose.

Advertisers, who had not noticed blogs before 2004, have responded in a big way. Understandably, with this much readership, they have started placing paid ads on blog pages, just as they do with web pages. But that is not all. Advertisers have also started hiring writers to pose as consumers or product experts and place comments and stories on blog sites to either promote with praise or refute negative comments with contradictory "stories" and "experiences." Since bloggers are anonymous (they write under a pseudonym, often just their email address) it is difficult to separate the advertising shills from the legitimate, independent bloggers. Because of the tendency to think of blogs as a forum for independent information, this makes the ad messages all the more effective. Some blog sites are sponsored entirely by a single company, which purports to conduct an "open forum" on its products and services but which naturally has its own stories to tell. An example is www.jalopnik.com a popular blog site that targets automotive enthusiasts but is sponsored by Audi, an automobile manufacturer. The Audi sponsorship is mentioned in small print on the front (landing) page of the blog site, where it can easily be overlooked. Also, blog readers who find information inside the site through keyword searches using Internet search engines may not learn about the sponsorship because their search bypasses the landing page. They may assume the site is independent and the information provided is objective, which it is not. Since blogs can be found using any search engine, a researcher may not even distinguish between a blog and a regular web site, which further confuses the issue of accuracy and reliability. Some bloggers are now calling for the development of a code of ethics to promote disclosure of sponsored vs. independent comments and stories.

The results of the early 2005 Pew survey, discussed previously, indicate that 62% of all searchers do not understand the difference between paid and unpaid (organic) search results. Only 18% reported that they were always able to identify paid ads regardless of location or format. Given the extent to which paid ads have become invisible (unidentified as such) this becomes more understandable.

The Search for Students

As part of the process of perfecting its business model, the search industry has become heavily involved with the academic community, both as a source of technology and also as a target for expanded business opportunities. While the original incursions on academic turf were coincidental to the larger goals of online search, the search business now conscientiously develops academic-like services. This comes at a time when students are flocking to search services as an alternative to traditional information sources such as libraries and also when non-profit institutions such as universities and public libraries are under increasing pressure to modify their financial model in the face of declining taxpayer support (I touch on the new financial realities for non-profits in Chapter 8.)

Take a moment to consider online search from the perspective of a University or School Board Member, CEO, CFO or other executive with a non-profit educational institution or community library. Imagine yourself in a budget meeting. You are sitting there struggling with yet another round of tough decisions when someone tells you that, instead of *you* having to spend your budget dollars to help your students access the web (which is something you think is educational because the web is expanding access to useful information at an exponential rate) somebody else will do it for you. At absolutely no cost! Free! No new money needed! This sounds good, the kind of help you could really use.

But wait, there's more. One of the leading search engines offers a "scholarly algorithm" to assist "serious" searchers as they look for "academically-oriented" information at no cost. Another one offers to scan your library's books for free and post them on their servers so they are accessible online to searchers. Surely these free services will help your students—indeed, all students-with their studies. By using these free services, demand for comparable services from the university or library goes down, thereby relieving budgetary stress (fewer people, buildings, furniture, books, periodicals, etc.) Who on the Board or in the administration can resist, especially when their students already like the service and use it constantly? The appeal to hard-pressed, under-funded academic institutions and libraries is simply irresistible. This is an offer they can't refuse.

In fact, the success of the search engine business model has resulted, in part, from the growing use of the search services by students. As more and more searchers use the sites, ad revenues go up. Searchers who become habitual search engine users as students are likely to continue habits learned as they move out into the professional world so that student use has both short-and long-term benefits for the search engines. (Targeting schools is a standard marketing strategy for many products and services, including technology, where Apple has led the way into the classroom.) How do they get more students to use the search sites? There are lots of ways; they offer free email accounts, for example, or maintain free emoticon libraries (I discuss the strange but enduring appeal of emoticons in the section in Chapter 9 entitled "A Future Without Words?") Yahoo does this and it draws them in. Also, it lends credibility and boosts traffic if colleges and universities praise and endorse the service or "partner" with one of the services, as some high profile non-profits have done with Google.

But there are even better ways. It is possible to advertise the service directly to students. This is how a vast array of youth-oriented products is sold, so why not search services? And, adapting the successful tactics of toy and junk food companies, the message is aimed at the youngest web users. The search engines are not waiting until students reach college age. Instead, the promo campaigns start in grade school or high school, when students are just learning about homework and term papers. Consider this Google logo, which appears all over the types of web sites used by middle and high school students:

This logo sends the message that it is OK to use Google's search engine to complete homework assignments. It grants approval, not only to the students, but also to any parents who happen to look at the web sites being visited by their kids. According to a recent Pew survey, 20% of all college students begin using computers between 5 and 8 years of age. One hundred percent (100%) of all college students have used the Internet by age 16, that is, before they get to college. So the habit of using search engines is well established by the time students reach campus as freshmen, which only adds to whatever justification universities need to accept search engines as a legitimate part of their academic regime. The fact that schools and universities get search tools and a whole lot of extra services for free from Google can't help but influence their inclination to see the positive aspects of online search using Google. Google makes money, the non-profits save money and the students gain access to information on a global scale. It is an unbeatable combination.

Not to be outdone, Yahoo has also launched a print media ad campaign aimed at students and their parents. It features Ben Stein, writer/actor/comedian (as he describes himself on his web site,) known for hilarious TV and movie roles such as the psycho college guidance counselor in the old *Charles In Charge* sitcom (in which he also played a sarcastic student loan officer at a local bank) and the economics teacher in *Ferris Bueller's Day Off*, recent author of a series of self-help books with titles such as *How to Ruin Your Life*, and game show host of Comedy Central's *Take Ben Stein's Money*. Mr. Stein is a former Nixon speechwriter and author of many online e-articles, with titles such as *Do Jews Run Hollywood?* and *I Love Wal-Mart*. The Yahoo ad shows him sitting in a very traditional looking library with an old-fashioned card catalog in the background together with a painting of Abe Lincoln and busts of Mark Twain and an unidentified Greek or Roman figure. Stein is wearing a t-shirt with the slogan *KNOW-IT-ALL ENGINE*. The accompanying ad text describes him an "actor and economist" (actually, it was his father Herb Stein who was the famous economist in the family but Ben did *play* an economist) and says he "uses Yahoo Search to appear more informed, erudite, and clever than any of his friends. He also uses Yahoo Messenger to gloat." The multiple messages here are not very subtle and the multiple audiences being targeted are not difficult to identify: libraries are outdated but Yahoo Search is "hip" (the band *Too Hip for the Room* recorded a song entitled "I Want To Be Ben Stein," which is available on the Internet as a free MP3 download.)

Information vs. Knowledge

The increasing involvement of search engine with students raises some obvious issues, especially in regard to growing advertising content, which is frequently unidentified. The Pew study shows that 68% of those surveyed believe that search engines provide "fair and unbiased" information, and that number increases to 72% for those under age 30. On the other hand, citing data for the 2nd Quarter of 2004 supplied by comScorenetworks, a marketing research company, Pew notes that between 40% and 45% of all search queries produce a list of hits that includes sponsored advertising. Pew concludes that, based on a comparison of these results, search engine users are "naïve."

In this context, we need to once again keep in mind that the purpose of advertising and advocacy is not to be "fair and unbiased" or to provide balanced or comprehensive information. Few ads actually claim to be objective. Ad information is usually not reviewed by impartial experts or required to meet standards for accuracy or analysis. (The Federal Communications Commission—FCC-occasionally bans an ad that contains blatant lies or other material that falls outside of community standards, but this is a rare occurrence.) *The purpose of advertising is to get the reader to draw conclusions and make decisions and to do so quickly, preferably impulsively.* As any sales person knows, too much detailed, objective information can delay a decision, sometimes indefinitely. Carefully circumscribed amounts of information are more effective in triggering a decision. This is typically what advertising or advocacy provides—limited information that has been carefully selected and then processed in such as way as to support a predetermined point, product or position. By influencing web page readers to draw conclusions and/or make decisions based on limited information, advertising turns information into false knowledge. The readers feel that they have enough information to make a knowledgeable decision or arrive at a correct conclusion when in fact they may not.

The inability to identify commercial content or even to recognize advocacy sites can become a liability for students, especially when overt advertising crosses over into subtler advocacy. For example, if a student is doing a term paper on energy resources and the environment and wants to learn more about petroleum-based fuels and their potential to pollute, they might find their way to the Exxon-Mobil web site (among others,) possibly with the help of a search query. There they would find information on pollution (supplied by Exxon-Mobil) conveniently placed alongside of information on Exxon-Mobil's petroleum-based products (for example, see the article entitled "Exxon-Mobile in the UK-Environment" dated January 20, 2005, on www2.exxonmobil.com/UK-English/Responsibility/UK_CR_Environment.asp).
But the information provided by Exxon-Mobil on the interrelationship of their products with the environment is not reviewed by outside industry experts or subjected to scrutiny or criticism by anyone with an alternate viewpoint. Exxon-Mobil does not provide a forum for contradictory information or alternate points of view. The company advocates a certain position on the impact of petrochemicals on the environment and is understandably selective in the information content it provides to support that position. The goal of Exxon-Mobile's web site is not to sell any specific product but rather to influence the way people feel about petroleum products. They attempt to provide knowledge about the environment, not just information about petroleum or pollutants. Of course they also know that building positive brand identification will eventually help them sell more gasoline and other retail products, but this is not their short-term objective.

The Exxon-Mobil example is important to an understanding the influence of search-based advertising on students because it reminds us that there is a difference between *information* and *knowledge*. Students may not understand this. Information can be thought of as the raw material — facts gathered that might be unorganized or even unrelated — used to construct knowledge. Knowledge can be thought of as an organized body of information that imparts comprehension and understanding. Building knowledge requires the processing of information by both individuals and groups. Processing involves such things as judgment, insight, understanding, remembering, wanting, sorting, choosing, comparing, interpreting, manipulating, reorganizing, evaluating and developing preferences for certain pieces of information based on their perceived value and accuracy as verified by alternate sources. Knowledge frequently is gained in the context of groups or organizations (employers, for example,) institutions (educational, as the prime example,) communities, networks (digital but also social) and society at large. Students, working alone (outside of a structured academic environment) and using search engines to acquire information that may contain advocacy or advertising bias or outright misinformation, either inadvertent or deliberate, are at a disadvantage unless they understand the need for and take the time to process information into knowledge. And an important by-product of the processing information into knowledge is critical thinking, the ability to take all of these evaluative skills and apply them to new or different batches of information.

Distinguishing information from knowledge is not merely a matter of semantics. Increasingly, these terms are used interchangeably, or at least confused with each other. The search engines are the beneficiaries of this sloppy thinking because, increasingly, they are said to be delivering "knowledge" to the desktop. They are not. They are bringing "information." Information is useful but knowledge is far more complex and, with guidance, leads to something that is even more complex-critical thinking.

Traditionally, students have gathered information that meets the criteria of accuracy and balance from a narrow range of sources such as the library, textbooks or scholarly monographs. These sources earn trust, in part by providing information in which commercial, political, philosophical and similar biases are either filtered out or identified more clearly (often with the help of a teacher, librarian or parent.) Most textbooks strive to be factually accurate and analytically balanced, sometimes to the point of having no position at all. Even in the digital age, most students grow up with and come to accept academic textbooks as the norm. Textbook content is vetted during preparation. Textbooks are withheld until they are reviewed by public committees and found to meet certain standards. Even after a textbook is released, it is monitored to gauge its impact. Still, with all this pre-publication scrutiny, textbook content is controversial on occasion; facts and analysis must be carefully researched assembled and then sometimes reworked in order to be acceptable to a sufficiently broad cross-section of educators and parents. Scholarly monographs usually undergo a similar process except that peers with established expertise and/or credentials in the appropriate academic field conduct the review. But due to advertising and advocacy influence on both access and content, this review process does not apply to a much of the information students get from the Internet when using search engines. And, because students have direct, personal access to information through search engines with no teacher or librarian involved, they are often the sole arbiter of what is good or bad information. For this reason, students may acquire information that they mistake for knowledge. The result is what we could call "false knowledge," questionable conclusions about or analysis of a set of facts caused by insufficient, incomplete or inaccurate information, or by subjective or otherwise flawed processing of information.

So, the opportunity to confuse information with knowledge has been increased by the phenomenal success of Internet search engines. This has been augmented by an increasingly common tendency to speak of one (information) as if it were the other (knowledge,) even by people who know better. This is not good, because not everybody knows better. An example, taken from the writings of someone (whom I will refer to as *Mr. X*) who certainly knows better, will serve to illustrate.

Mr. X is a well-respected author, journalist (*New York Times Magazine*, *Wired Magazine*) and computer expert. The first book he ever wrote was about the computer hacker phenomenon. His latest book is a history of cryptography. These are not simple subjects. He popularizes science and technology and defends them against their detractors with great enthusiasm. In this sense he reminds me a little of the late Carl Sagan. His claims to fame include—literally—finding Einstein's brain in two mason jars in New Jersey in 1978.

Mr. X is the Chief Technology Writer for one of the largest U.S. news magazines and, as such commands a weekly readership of hundreds of thousands if not more, including (no doubt) many high school students and their parents and teachers. He has influence over what people think and believe about technology. In the December 27, 2004 issue he wrote an article about Google and the spectacular effect ("life-changing feedback that comes from a query in the Google search field" as he describes it) search engines have had on the future prospects for, among other things, the truly digital library, the excellent prospects for which he attributes to the efforts of Google (with a nod also to Amazon's A9, which I discuss in the section on "Book Search the Amazon Way" in Chapter 3.) Here is a quote from the final paragraph.

"...there is the very big issue of how much we want the world's *information* transformed into a giant ad environment. But in this case *knowledge* and commercialism seems inextricably intertwined." [Italics mine.}

What we see here is an increasingly common instance of *information* morphing into *knowledge*, in this case in just one sentence and by someone who certainly should understand the difference. The irony is that he was touching on the possibly pernicious influence of advertising on the usefulness of Internet search engines (that is why I chose this particular example.) But this unfortunate interchange of terms, which in Mr. X's case is probably just a slip of the pen, propels him (and his many readers) down the slippery slope. Schuss! Information becomes knowledge. It is not. Search engines do not dispense knowledge.

I am sure that many students and other web users recognize advertising influences when they see them (at least 30%, according to Pew) just as I am sure a certain percentage of students have the know-how to bypass web pages and mine databases and other online resources less influenced by advertising. But a great many do not and the invisible ad tactics developed by commercial digital information suppliers (aimed at increasingly younger users) have never before been encountered on such as massive scale. Search engines are unlikely to be held responsible for this. They can (justifiably, in my opinion) maintain that the burden of responsibility to gather complete information and then draw proper conclusions falls on the user, not on them. Their role is merely to provide links to information whether it is valid or invalid, complete or incomplete. It is the search engine user's responsibility to understand and maintain the standards that differentiate information from knowledge. It seems to me that if the academic community permits and even endorses the use of search engines (and how can they refuse-why would they?) then they have a responsibility to nurture, publicize and also, if necessary, enforce these standards. This poses a challenge for educators but create a major opportunity for libraries.

Information About Information

Because information from the Internet is replacing information from traditional sources, there is a need to educate students and the community at large about digital content and the differences between it and more traditional content from libraries, textbooks, periodicals, monographs, etc. Libraries are in a unique position to serve as an advocate for careful user evaluation of information and for an understanding of the differences between information and knowledge. These are serious challenges to both the integrity of content (which students used to be more able to take for granted) and to the goal of helping library customers improve their capacity for critical thinking (which depends upon knowledge gained from a broad spectrum of reliable information.) To this end, libraries can serve as a catalyst, by promoting ongoing discussion of this and related issue (seminars, guest lectures, etc.) creating online tutorials to help library customers learn about search engine strategies and tactics (for example, www.lib.berkeley.edu/TeachingLib/Guides/Internet/,) evaluating information obtained from search engines and researching search engine evaluation web sites (such as www.imediaconnection.com or www.searchenginewatch.com) and drawing interested administrators, faculty members and community resources (experts in advertising, law, ethics, epistemology, economics, information technology, and library science) into the process as guest speakers, bloggers or as guests or moderators in live chat sessions. Libraries can also use their buzz marketing programs to create awareness, foment debate, direct customers to ad-free content, etc. Libraries have always been a critical source of information; now they can become a source of information *about* information, especially as it pertains to the Internet search phenomenon. Students need this immediately. This is a realistic, achievable goal for all libraries and librarians. It should be written into the mission statement of every library.

6. Ambient Information

Ambience is defined as "that which surrounds or encompasses as it pertains to environment, atmosphere or milieu." You may think of ambience more in terms of cutting-edge interior design in a trendy restaurant or upscale boutique, where light and color, unusual furniture, artwork and unique architectural details are used to enhance the dining or shopping experience of patrons. But ambience can also describe other things that surround us in our daily lives. For example, we have already experienced a long-term trend toward "ambient entertainment." Music "background" has slowly but surely filled every available niche, including restaurants, automobiles, dentists' offices, malls, even elevators. Music promoters load tunes onto iPods and rent them to restaurants, nightspots, clothing boutiques, jewelers and hair salons, who then loan them to patrons and shoppers while they are in the store. Joggers wear a Walkman or iPod when they exercise. No place is off-limits; a popular gift item in recent years has been waterproof radios for use in the shower.

Much of this ambient audio has acquired an advertising component, which pays for the entertainment portion of the programming. Ambient advertising can be subtle. Serious attention has been given to the use of "motivation music" (called Stimulus Progression) in consumer and work environments. The idea is that, above and beyond any overt advertising messages that accompany the music, the right musical background can trigger customer interest and buying responses and can even cause shoppers to linger longer in a particular store, increasing the probability that they will spontaneously purchase something. One recent study showed, for example, that classical (as compared with pop) music playing in the background at restaurants increased the amount of food purchased and also the size of the tip given to servers. Beethoven, Mahler and Vivaldi seemed to have the greatest effect. The least money was spent when there was no musical background at all. McMozart anyone? (On the darker side, 7-Eleven stores play classical in their parking lots because it reduces loitering and related crime.) Offices and retail stores use background sound tracks to calm or energize employees and customers, depending upon the time of day — energize in the early morning, calm during coffee breaks and lunch, then high energy again to start the afternoon and end the day. Music is not the only ambient audio used to motivate or energize. Some companies take a more direct approach. During the dot-com boom, tech companies piped business news and stock reports into offices and cubicles to motivate employees; cheering often accompanied an uptick in the company's share price for the day.

More recently, music and advertising are being augmented by news and other, more information-oriented content. Flat panel screens now make video news (and also entertainment,) together with the usual component of advertising, available almost anywhere. Sure enough, we can "watch TV" in sports bars, airports and airplanes, restaurants, Wal-Mart, bank lobbies, doctors' waiting rooms, in our SUV's and in every room in our house including the bathroom. Even elevators, formerly the exclusive domain of audio, now have video, which brings news, sports scores and stock reports to busy workers as they move from floor to floor. For those wanting to fight back, a device called TV-B-Gone, cleverly disguised as a key fob, holds infrared codes for most TV's and enables the user to turn sets off in public places.

As content has expanded to include information, the idea of "ambience" has taken on new meaning. It is now used to describe the constantly changing background information that we use throughout the day. Examples include not only news and stock prices but also weather, local traffic, sports scores, airport schedule changes, health indices such as pollen counts and UV levels, local and regional crime and missing persons alerts, homeland security and other emergency bulletins, etc. What makes this type of information unique is the fact that much of it changes so frequently that we have little interest in the specific data at any given moment but instead are more concerned about the data *trends*. Is it getting warmer or cooler outside? Is the wind going to pick up or die down this afternoon? Is the market up or down in the last hour? Is the dew burning off so we can mow the lawn? Is traffic flowing normally on the Interstate or is there unusual rush-hour congestion? Are we at Code Yellow or Red? What is the best time today for the fish to bite? Normally, this information is accessed the same way any other type of digital information is acquired, using a PC, laptop or PDA that displays alpha/numeric data. In analog format, we typically acquire information of this sort from TV or news radio. The data search has to be initiated by the recipient (turn on the computer, select the news radio station) and then read or heard in detail even though the actual alpha/numeric information has little or no "shelf life" because it is constantly changing. We have to "check" continuously every few minutes or hours to establish the trends as they evolve. It requires constant activity and effort to remain current.

But, as nationwide Wi-Fi networks are deployed, ambient information can be broadcast continuously for automatic display on compatible digital devices such as the *Ambient Orb* (www.ambientdevices.com) a glass ball that allows us to simply glance at color changes to "read" the latest trends. We don't need to know the exact market price of our favorite stock because it will change in a few minutes anyway. What we DO want to know is whether it is trending upward or downward as compared with yesterday's close. Since we don't need exact alpha/numeric data, the Orb converts the variable information into colors, what the inventors of the device call a "glanceable thing." Other devices, such as the *Ambient Dashboard*, have replaceable faces so that many different types of information can be tracked by a single, unobtrusive device that can be placed anywhere in the home or office, even in a briefcase, ready to be viewed at a glance.

There are many advantages to converting ambient information into colors or trend lines, including the speed with which the viewer can determine the trend (instantaneously) and the absence of the mental clutter than comes with "learning" information that changes almost immediately and has to be "forgotten" to clear the mind for the next batch of temporary data. It is like glancing out the window to check on the weather; easy, quick and (at least for the moment) accurate but not a distraction that places much demand on one's analytical skills or memory capacity.

Color-coding is probably not the future of information distribution (although color coded alerts issued by the U.S. Department of Homeland Security has significantly increased public awareness of this method of communication.) But the commercial success of color-coded ambient devices indicates both an acceptance of and a demand for ambient information by the populace at large. We are becoming accustomed to being surrounded by information, whether audible, visual, color-coded, etc. We are learning that information has immediate value that can be maximized if it is instantly available, regardless of where we happen to be at the moment. So, when considering ambient information, why stop with information that changes frequently? All information, even information that never changes, can be ambient if by that we mean readily available anytime, anywhere through a wide variety of wireless devices. When this happens, consumers will be able to access product information (specifications, independent reviews, price comparisons) right in the store while they are shopping. Students will be able to check historical facts or review a particularly relevant paragraph from a book spontaneously, during a bull session in the student union, for example, when a "learning moment" is at its peak. Doctors and their patients will be able to check medical records from the office, hospital or home. An attorney will be able to review case law on any subject while commuting to and from the office. This type of information is not "glanceable," it is much more permanent and the same data may need to be accessed multiple times over an extended period. But there is no reason why it cannot be ambient and readily accessible for either quick or leisurely access, whichever is appropriate. Since a lot of mobile professionals use a Personal Digital Assistant (PDA) when they are away from their office, ICANN (The Internet Corporation for Assigned Names and Numbers) has recently approved .mobi as a new domain name for web sites. Mobile technology companies like Nokia and T-Mobile as well as Microsoft sponsored the request. The idea is that users would know that web sites with the .mobi suffix (http://www.whatever.mobi/) would be specially designed to work around the limitations of cell phones and PDAs, which have tiny screens and limited data storage capability. This will encourage even more users to view web pages or alpha/numeric information on their wireless devices.

The only thing stopping ambient information is the lack of ambient access. The answer to that problem is Wi-Fi.

Wi-Fi

On October 22, 2004 the Mayor of San Francisco announced that the City has set a goal of free, wireless broadband Internet access for every resident. Philadelphia's mayor has announced that all 135 square miles of the city will have high-speed wireless Internet access by summer of 2006. This type of service is referred to as Wi-Fi, a collective term for several different and competing wireless protocols. "No San Franciscan should be without a computer and a broadband connection," that city's Mayor said in his announcement. (Officials noted that the stadium where the San Francisco Giants baseball team plays already has Wi-Fi, presumably so that fans can be at the office and the ball game at the same time.) Philadelphia already has Wi-Fi at their Convention Center, the Reading Terminal Market, JFK Plaza and upscale Rittenhouse Square area. "We looked at it as a way to be a city, literally, of the 21st century," says a spokesperson for the Mayor's office. "We want to bridge the digital divide for residents who wouldn't [otherwise] have access to the Internet, particularly schoolchildren." Plus, the service could help make Philadelphia "hip" enough to stem the outward flow of college graduates, a major concern for the City.

Other cities from San Jose and Long Beach in California to Corpus Christi, TX, Chaska, MN, St. Cloud, FL, Phoenix, AZ and Atlanta, GA have made similar Wi-Fi announcements. Major cities such as New York and Dallas say they will follow suit. They join growing list of cities and towns across North America that have committed to installing wireless broadband networks. City-wide Wi-Fi has already been deployed or is on the agenda in many small towns and rural areas where there are concerns that business and residential development will bypass them if they cannot provide high-speed Internet access. Even major cities feel the competitive pressure. Atlanta's FastPass network is intended to improve the city's ability to attract both tourists and business relocations away from cities that are promoting themselves as more "Wi-Fi friendly." In Iowa and West Virginia, two states that have been bypassed by large-scale fiber networks, non-profit corporations such as OpportunityIowa and iTown Communications are installing combination Wi-Fi/ fiber networks to get residents connected. Along with municipalities, they now speak of broadband the same way they speak of public utilities like water or sewer services. "It's becoming an expectation for customers," says Mike Landman of Atlanta, "like having a bathroom."

The "Broadband Gap"

What is going on here? Why does the San Francisco city government have to step in, in the land of Silicon Valley no less, to bring broadband to the masses? Why is Philadelphia concerned about lack of Wi-Fi in its Business District? Why are rural areas so interested in this new technology? The answer is that while the rest of the world has rushed to install broadband communications technologies the United States has lagged, and the consensus is that it is hurting our economy. According to official Federal Communications Commission (FCC) statistics, updated in September of 2004, the U.S. ranks either eleventh (11th) or twelfth (12th) in the world in broadband deployment, depending upon whose surveys you believe. There are more subscribers in the U.S. (32.5 million versus 24 million in Europe and 14 million in Japan) but on a per capita basis—subscribers per 100 people-the United States with its 6.9 subscribers per 100 residents ranks behind 10 other countries. South Korea is the global leader with 21.3 subscribers per 100 people, followed by Hong Kong (14.9), Canada (11.2), Taiwan (9.4) and Iceland (8.4).

Explanations for the U.S. broadband gap vary, but the most popular one is that many of the countries ahead of the U.S. have more concentrated populations while we are spread out over a much larger area, making it prohibitively expensive to install the necessary wire or fiber optic cable. And yet Canada, a huge country with sparse population, has twice the wired broadband connectivity as the U.S. The FCC even denies there is a major problem, and in their latest bi-annual report to Congress (mandated by the Telecommunications Act of 1996,) concludes,"We have indeed turned the corner on digital migration." The U.S. government's goal is for "universal and affordable broadband access" by 2007. They have a long way to go according to statistics released in late 2004 by the Department of Commerce. Only 19.9% of total U.S. households have broadband and one third of America (including me) still accesses the Internet via 28.8 kilobits per second (Kbps) dial-up. By the end of 2004 53% of residences that use the web had broadband according the Nielsen Ratings but 14.2% of households have yet to go online. About 70% of households with broadband access go online daily as compared with about 50% of those with dial-up access. Broadband users stay online an average of 107 minutes each session while dial-up users stay online an average of 86 minutes.

Despite FCC assurances, individuals, business and cities all across America continue to be critical of U.S. policy, citing our weak and eroding position relative to global competitors. Even the official benchmark for broadband speed used by the FCC, 200 Kbps and then only in one direction, is subject to ridicule. In many of the countries ahead of the U.S. in broadband deployment, 200 Kbps does not even qualify as broadband. In Japan, for example, the broadband benchmark speed is 8,000 Kbps, which is currently unattainable commercially anywhere in the U.S. Some question whether the U.S. currently has any broadband at all by global standards. Critics also site the exorbitant cost of broadband in the U.S. The average monthly broadband subscriber rate in Japan is $10.00. It is a minimum of three times that in the U.S.

The Information Politics of Telcos

Major reasons cited for the slow deployment of wired broadband, in addition to geography, include the oligopolistic nature of the U.S. telecom industry and public policy decisions designed to balance the preservation of jobs with the deployment of new technology. Throughout America's wired history, a few major companies have dominated the markets. Bell Telephone and AT&T controlled local and long distance telecom for decades, installing and maintaining hundreds of thousands of miles of wire and paying for it with ever-rising local and long-distance rates. More recently, installation and control of the wires running into homes and businesses have been turned over to the remnants of Bell ("Baby Bells" such as Verizon, SBC and Bellsouth Corp. and Qwest Communications) and upstart "cable TV" companies like Comcast and Adelphia. FCC regulatory policies, claiming to seek balance, tend to swing pendulum-like from promoting competition in order to lower consumer costs and improve services to reversing many of these decisions under pressure from various political and business interests. For example, the FCC encouraged local phone companies like the Baby

Bells to offer long distance services while simultaneously allowing long distance companies like AT&T and MCI to diversify into the local phone service business. More recently the FCC has reversed these policies, the result being that many companies have exited the local phone service business, thereby reducing competition.

One reason for recent FCC policy reversals is that the Bells have promised to "rewire America" and bring fiber-optic cable to millions of additional homes and business over the next several years, investing billions of dollars and creating thousands of new jobs in the process. The costs of bringing fiber to an individual home can range from $300 to $2,000 and in order to encourage the Bells to invest this much money the FCC has promised that they won't have to lease their excess capacity to third-party competitors at a discount. In reading the fine print, however, we discover that only Verizon has actually committed to bring fiber the "last mile" (all the way into the home) and Verizon is asking for more concessions to protect its monopoly before it makes the required investment. Bellsouth will bring fiber to within 500 feet and SBC to within 5000 feet. They will then use existing copper wires to bring the service the rest of the way into the home, presumably to cut installation costs. Copper has severe bandwidth (speed and capacity) limitations, however, and will not even come close to true broadband as measured by global standards. We will all be wired, but for sub-broadband service. Of all the Bells, only Qwest is limiting its commitment to more wire and fiber optic cable, indicating that it will install fiber in new houses but will service existing homes with wireless data transmission, which is much less expensive to install.

Regional and long-distance phone companies, who sell broadband Internet to consumers and businesses, have intensified a national campaign to quash municipal wireless initiatives. Qwest, Sprint, Bellsouth and Verizon have pressed for legislation in Florida, Utah, Louisiana and Pennsylvania. A dozen states already regulate public-sector telecommunications projects or ban it outright. The U.S. Supreme Court has upheld some of these laws. The telcos consider municipal Wi-Fi unfair competition and, along with other new technologies such as Voice Over Internet Protocol (VOIP,) a threat to their business model. In Pennsylvania, Verizon actively supported House Bill 30 that prohibited "political subdivisions" from "competing" with private telcos by offering municipal Wi-Fi or other communications services. The Bill also authorized an additional $3.0 *billion* dollars in rate hikes to help pay for Verizon's fiber network expansion. The legislation was targeted primarily at the City of Philadelphia Wi-Fi program, which, under the proposed legislation, would have had to be operational by 2006 to receive state approval. But in order to get the legislation signed by Governor Ed Rendell (former mayor of Philadelphia,) Verizon had to reach an accommodation with the City of Philadelphia specifically exempting the city from the legislation. This was described as a "win/win" compromise between Verizon and the city, but at the expense of broadband connectivity for everybody else in the state. All other municipalities in Pennsylvania are prohibited from developing municipal broadband services of any kind, so that other large cities like Pittsburgh will be completely dependant upon Verizon for service. Considering the huge competitive advantage this gives Philadelphia, many more lawsuits are anticipated. Various other approaches are being worked out to protect municipal Wi-Fi networks from the wrath of the big telcos. The city of Atlanta, for example, has partnered with a private firm to provide a defense against Philadelphia-like lawsuits from broadband companies servicing the metro Atlanta area.

Apparently, even the FCC has doubts about the quality and cost-effectiveness of the "rewire America" program. In October, 2004 the FCC said it was ready to give the go-ahead to new rules that let consumers receive broadband through the 120VAC outlets in their homes (called BPL — broadband over power line.) Electricity utilizes low frequencies, so that that there is plenty of room at higher frequencies for data transmission and telecom. Utilities already use their own power lines to transmit data on electrical grid performance. Power utilities say that broadband could be brought directly into the home and accessed through special modems that would plug into the standard electrical receptacles already installed in 99% of all homes in America. Or, broadband could be carried over the electrical grid to Wi-Fi antennas mounted on utility poles and then broadcast into homes and businesses. The FCC has clearly indicated that this is in the public interest as an alternative to cable. Southern Company, one of the largest utility holding companies in the U.S., has announced that it is running technical trials and a market test in Alabama. What is being described as a "pilot program" is also underway in Cincinnati, OH. Consolidated Edison in the Northeast and Hawaiian Electric Company are also running beta programs. Broader use of the electrical power grid's bandwidth could also lead to standard phone service being bundled with electrical service by utilities. In should be noted that many communities in the U.S. provide electrical service via municipal utilities. They may soon be offering broadband and wired telco over the same wires.

Cable TV companies have been especially fortunate recipients of favorable FCC rulings. These have had the effect of minimizing competitive pressures, much to the displeasure of consumers who complain constantly at Congressional and FCC hearings about high rates and bad service. Predictably, a large cottage industry of cable signal "decoders" and other illegal devices have proliferated as consumers try to circumvent what they claim is price gouging. The probable explanation for lenient FCC regulatory policy toward cable service providers is the extremely high labor and materials cost of laying and then maintaining cable to each individual home in a given geographic area.

There are also public policy issues that have little to do with broadband connectivity. An important one is that oligopolies are slower to shed jobs in times of economic downturn. The financial collapse of MCI and the subsequent loss of many tens of thousands of high-paying jobs have influenced FCC thinking about the wisdom of unfettered market competition in a time of uncertainty and technological upheaval. Telecom as a whole lost 40,000 jobs in 2003 and that number was expected to be higher in 2004. Wired connectivity, however, is labor intensive both for initial installation and ongoing maintenance. Wireless is the opposite. Once installed, experience with cellular phone service wireless networks has shown that they require far less ongoing maintenance. The benefits of wireless accrue to the *users* of the service more than to the *providers*. But maybe, the FCC reasons, it is better to go a little slower, even if the consumer's monthly cable or phone bill is a bit higher. The fact that many of these policy decisions were made in 2004, a re-election year, cannot be discounted.

The high cost and unpredictable service provided by cable companies has become a larger issue now that cable companies are "bundling" Internet service with entertainment services. Comcast is a prime example of cable struggling to meet the needs of Internet users. Restrictions on competition have given the company pricing power and, because entertainment revenues are more important than Internet access, Comcast limits cable Internet to residences only, refusing orders for business installations. Comcast also limits the number of outbound email transmissions per day (to reduce network load) and places restrictions on other services normally provided by Internet Service Providers (ISP's.) Cable Internet subscribers share bandwidth with entertainment and commonly experience severe slowdowns and even complete network disruptions during periods of peak TV usage. This makes home-office use of cable Internet unreliable. One positive side effect of high cable rates is that satellite television (DirectTV, Dish Network, etc.,), which is expensive to deploy, has prospered. Satellite rates can approach $100.00 per month for entertainment services only. Satellite Internet access is very expensive, especially for bi-directional services.

Local Area Private Wireless Broadband Networks

Municipalities are not the first to discover the advantages of Wi-Fi as an alternative to wired services. For the past couple of years private businesses have been setting up local area Wi-Fi "hot spots" to lure customers and their laptops. Print shops (Kinko's,) fast food restaurants (McDonald's,) bookstores (Borders) and coffee houses (Cosi) have promoted these services at no-charge. Almost all of the major hotel chains from the Ritz-Carlton to the Holiday Inn Express are also adding hot spots for the convenience of the business crowd. Other areas such as airport lounges have hot spots and charge a fee for travelers to log on and work during layovers between flights. Computer clubs set up hot spots for their members and a few public-spirited digital citizens have even set them up as a service to the neighborhood where they live. The State of Texas Parks and Wildlife Department has announced that five of their state parks will offer wireless broadband starting Jan. 1, 2005. Texas is also installing Wi-Fi at rest stops along State highways. California has contracted with SBC to provide wireless Internet access in 85 state parks, but service will only be available for a fee of $7.95 per day. Many areas are dotted with small hot spots but there is usually no pattern. Potential users have to drive around until they detect a hot spot signal or they can consult one of several hot spot locator web sites. These directories, such as www.wi-fihotspotlist.com/, list public hot spots by zip code for the convenience of users. Most large cities have a few or even several hundred spots and the number is growing steadily. Of course there are fewer in outlying areas, but even small towns are joining the network. Madison, GA, population 3,600, which is the largest town in my county, has a hot spot—*Amici's Italian Cafe*. Directory service Jiwire, Inc. estimates that there were about 22,000 hot spots in the U.S. at the end of 2004.

As hotspots spread, consumer products are starting to appear on the market that depend upon Wi-Fi access from almost anywhere. Just one example is a digital camera introduced by Kodak in 2005. Photos can be emailed directly from the camera using wireless Internet technology from any available hotspot. They can be sent to a personal PC or they can be uploaded directly to Kodak's Ofoto web site, where anybody can view them and purchase prints. The camera will also enable users to download and view photos previously stored on the Ofoto site. To launch the new product, Kodak has teamed with wireless service provider T-Mobile to provide the wireless connectivity at no cost. It doesn't get any cheaper than free.

Not all Wi-Fi hot spots are small in scale. The 2.5 million-square-foot Time Warner Center being built in Manhattan has taken a novel approach—Wi-Fi as a utility service (like electricity) that can be turned on and off by users who are billed for usage on a metered basis. Wireless Internet access is guaranteed anywhere in the complex. So is cell phone service and two-way radio communications for employees. Called "utility computing," pay-as-you-go wireless is expected to become a standard feature of Class A office buildings in the near future. Atlanta's Hartsfield-Jackson International Airport, one of the largest in the world, will be completely blanketed with Wi-Fi by March 2005. The Atlanta access rate will be somewhere between $3 to $9 dollars for unlimited usage for a day.

Wi-Fi technology is extremely affordable even for individuals and small businesses. Companies like Proxim offer hot spot starter kits for about $250.00, which they claim can be installed by anyone with reasonable computer skills. Wi-Max is more expensive but can cover a much larger area than standard Wi-Fi, up to 30 miles from a single antenna as compared with a few hundred feet for Wi-Fi. PC manufacturers like Dell and Hewlett-Packard include wireless software and built-in antennas as a low-or even a no-cost option on most desktops and laptops. Wireless will also get a boost from SONY, who in early 2005 introduced a new game console called the PlayStation Portable (PSP), which allows gamers to connect with the Internet to play with other gamers anywhere in the world. To appeal to adults as well as children, the PSP will also be able to download music, photos, movies and television shows from your home network. If all that isn't enough, PSP can be used as a voice over Internet phone (VOIP.) There are well over 100 *million* gamers worldwide (mostly of student age) and by early next year they will all be connected wirelessly. And finally, after a rough start, wireless security is now on a par with wired networks. Based on the universal acceptance of and user satisfaction with wireless phones over the past few years and the immediate prospect of millions of new wireless gamers becoming accustomed to complete mobility, there appears to be no significant user barriers to Wi-Fi, except access.

What municipalities are concerned about are the non-continuous zones of coverage ("coldspots") by the small-scale, private and for-profit hot spot movement. Without total coverage neither businesses nor individuals can be sure of high-speed access. They are also concerned about the apparent technical limitations and high user costs built into the wired networks, both existing and planned. Wireless is not only far less expensive for initial installation, it is also easier to upgrade as technology improves ("future-proof," in industry terminology.) The City of Philadelphia, for example, estimates that their complete Wi-Fi network will cost only $10 million to install and $1.5 million per year to maintain, more than small-scale business networks but only a tiny fraction of what the large wired networks will cost. Tucson, AZ estimates their installation costs to be between $100,000 and $150,000 per square mile. Wireless can also be installed faster than wire, leapfrogging the fiber (or copper) lines directly to individual PC's and providing even higher rates of data transmission. In fact, high-speed wireless can use the same CDMA (Code-Division Multiple Access) network that is already in place to route cell phone traffic. Verizon, for example, has introduced BROADBANDAccess, a wireless Internet service that delivers 300—500 Kbps (with bursts up to 2000Kbps) over the same network used by Verizon cell phone service, which already covers most of the continental U.S. Most areas of the country may already be "wired" for wireless.

Music Technology

Music technology merits special attention. Online music and video downloading with peer-to-peer (P2P) software is one of the Internet's greatest technical success stories, an application that is so popular that it commands, at any one moment in time, an estimated one-third of all available bandwidth in the world. Downloader pioneers have become cultural icons. But downloading is controversial. In just the past couple of years (since 2003) it has become a top priority for the Recording Industry Association of America (RIAA,) the Motion Picture Association of America (MPAA) and the lawyers who specialize in intellectual property matters for the entertainment industry. Downloading has become a high-profile legislative and judicial issue, with constitutional implications. It has spawned new marketing strategies and economic theories, and it has raised serious moral and ethical issues. Everybody has a perspective. Music users rail against price rip-offs. Economists call for new business models. The courts are split. Established artists protect their turf against new acts. The newbies fight back by launching free MP3s and film clips into cyberspace. Technologists develop code to protect property rights (a/k/a copyrights.) Other technologists defeat this code in the name of "commons rights." It sometimes seems that lawyers just want to keep the controversy going and growing. Fans are passionate about music (and also about some vague notion of online freedom) and the entertainment industry is passionate about their business model, which has made everybody from artists to roadies very rich. This is a quintessential conflict between the establishment and progress, the old and the new. Balance has not yet been achieved.

Libraries (and their universities, if they are associated) have been slow to respond to the issue of music downloading, understandably cautious because of threats concerning litigation (I discuss music download litigation later in this Chapter.) The risks of getting involved seem to outweigh any possible benefits. But it may take years, even decades, for "copyright" or "digital commons" issues to be resolved in the courts. Moreover, it is not necessary to have these legal issues settled in order to embrace the digital music movement. There are important benefits to doing this. It is time for libraries face the music.

The music debate in the 21st century is not really about music at all. It is about technology. Too many post-millennials (the term millennial refers to today's student generation and so I'll define post-millennials as people over age 30) don't seem to understand this. They frame the issue in the same way it has been framed since the 1920's when "flappers" and jazz scandalized the establishment. Now as then, tsk-tsking about semi-nekkid dancers and raunchy lyrics is widespread. I do not mean to ridicule the moral and ethical issues raised by popular music. But, like the legal debate, the moral debate about music, which cuts across many decades and music genres, is in a constant state of flux. It is not going to be settled any time soon. Besides, most music pundits don't really expect today's "hip-hop" to last any longer than yesterday's "grunge." Something even more outlandish will replace rap one of these days—it is the nature of popular entertainment to inflict temporary shocks to the establishment. It makes money and is apparently lots of fun and it explains much about the creative energy of many entertainers.

Post-millennials tend to view music in moral, ethical and, with a nudge from RIAA lawyers, legal terms. They think that "theft" of music and movies (along with nudity and foul language on TV,) if left unpunished, adds to what is already a perceived decline in moral standards. This is a message that resonates. It is logical from there to attack the technologies especially peer-to-peer (P2P) but also cable TV that facilitate the decay. Framing the debate in terms of morality or ethics deflects attention from a much more complex and self-serving agenda favored by the RIAA and others, that being the perpetuation of tight control over music production and distribution and the profits that flow from that control. That this works is just another indicator of the extent to which post-millennials have failed to come to grips with the digital world.

Its Only Rock and Roll but We Like It

When I was a teenager, it was normal for my elders to frown on Elvis' swivel-hips, the Beatles' long hair, face makeup on Kiss, etc. Rock and roll made adults nervous. In one of the most incredible TV shows ever broadcast (in 1956, which as a child I saw live on a small black-and-white set) Ed Sullivan put his arm around Elvis (after a performance — I think it was "Jailhouse Rock" — that could be viewed only from the waist up because of the sexual innuendo in his dancing) and assure America that "he is a good boy." Sullivan did the same for the Beatles in a show broadcast Feb. 9, 1964. These appearances helped legitimize rock music. Sullivan reassured his audience and many felt better afterward knowing that "the kids" were not being corrupted. The record industry made millions. There was no technology agenda to complicate the issue.

In retrospect an interesting thing about the pre-digital era was the simplicity of music technology. Elvis performed accompanied by a slap bass, two guitars (one hollow-body,) a drummer and a gospel group. There were two microphones on stage, one for Elvis and the other for the backup singers. A decade later the Beatles used an electric base, two guitars, a drummer (Ringo!) and four microphones. Not much else had changed except that the Beatles wore suits. The Ramones used pretty much the same setup as late in the early 90's except that they played a lot faster. Throughout the 60's, 70's, 80's and even the early 90's, aside from the electric base, there was little or no change in performance technology. I remember the buzz generated when the mid-60's band The Lovin' Spoonful used what appeared to be an electric zither in a televised performance of "Do You Believe in Magic?" In the era of pre-digital rock and roll it qualifies as a big moment in music technical history. A new technology called a synthesizer (also referred to as a Moog synthesizer, after its inventor, Robert Moog.) was introduced in the mid-60's. It was a keyboard device with transistors and a few cathode ray tubes that played what came to be called "electronic-sounding" notes and chords. Synthesizers were used by innovative groups such as Emerson, Lake and Palmer, whose synthesized anti-war song "Lucky Man" was a milestone in new music. But the synthesizer was ignored by many of the biggest groups. Both the Beatles and Rolling Stones acquired synthesizers but never use them in recordings. Big-name musical acts were not enamored of cutting edge technology.

From a recording and distribution standpoint, records were sold on vinyl disks (albums and singles) all during this period. The 45 revolutions per minute (rpm) speed standard for 7 inch vinyl records was established in 1949, and 33 1/3 rpm for albums came about the same time. (The previous standard had been 78 rpm, so at slower speeds more music could be put in a disk and/or the disk could be smaller.) Other analog playback technologies emerged, including reel-to-reel tape decks that appeared in the 50's, audiocassettes in the early 60's, and 8-track cassettes in 1965. But vinyl records, recorded in sound studios onto magnetic tape (an analog process) and then mechanically pressed from metal master plates, remained the most popular distribution medium, in large part because everybody had record players. It was not until the late 80's that compact disks (CD's — a digital format) outsold vinyl records.

The crucial point to understand is that none of this recording or playback technology challenged established methods or channels of production or distribution or caused discomfort to the music industry's business model. The effects were limited to steady but not dramatic improvements in quality and/or convenience Eight-track made it possible to play albums in cars, reducing dependence on the radio. Radios retaliated with stereo FM. Audiocassettes made possible the SONY Walkman, which made music personal. The first Walkman, which we can think of as an analog iPod, was introduced in 1979. These incremental improvements were welcome, however, and music flourished largely due to the wider exposure bands received from aggressive marketing (live concert tours and TV appearances.) Record sales rose steadily from decade to decade. Retail outlets — everything from music stores to grocery stores — were the point of sale.

No Such Thing As a Copy

All of that seems laughable by today's standards. Music is no longer recorded onto analog tapes in sound studios using musicians and singers. The artists still participate in the process but it is more accurate to say that music is assembled digitally from pre-recorded "loops" — sound tracks — that may have been captured (sometimes in home studios or live events) weeks or even months prior. Many songs also contain material from previously recorded music, a process called "sampling." Since the overall result is blended together digitally using commercially available software such as SONY's Sound Forge 8, synchronization of tempos or musical keys is not a problem. Almost all music is digitally mastered and all distribution copies (primarily CD's) are made from digital masters. Digital technology has solved the quality problems that were only marginally improved upon over the course of the prior 30 years. In 1998 the MP3 (short for MPEG-3) format was introduced. MP3 is an exact digital replication of the original digital master. Moreover, the MP3 format is compressed to a reduced file size so that it takes up less storage space (on the hard drive of a PC, for example.) Better yet (or worse, if you are the RIAA,) MP3's (and similar formats) can be copied onto hard drives or from CD to CD or distributed quickly and easily over the Internet. There is no need to have a "master" of any type, either hard copy (CD, cassette, vinyl, etc.) or a digital master (such as the recording company uses to manufacture CD's.) *Every copy is a perfect original.* Small, handheld devices like Apple's iPod, which was first sold in 2001, can store thousands of these compressed but otherwise perfect digital files.

These innovations and improvements do not fully explain current conflicts over digital music. Another technological leap has occurred, brought about by something unrelated to music or musical technology. An embryonic "information commons" has come into being, thanks to the Internet and to peer-to-peer (P2P) file distribution and retrieval software. (I discuss P2P in more depth in the next section of this Chapter.) P2P software is free or nearly free, simple to use and can directly link thousands or even millions of individuals to each other simultaneously, through their personal computers. (Not even the most advanced search engine technology can presently do this, although they are trying.) P2P was not developed to enable music file sharing. It was developed to enable *any* kind of file sharing. Music file sharing is a byproduct of P2P, not the other way around. But right now millennials use it mostly for exchanging digital music files, and in doing so have made it one of the most widely used software programs in the world (arguably second only to Microsoft's operating systems.) Music is copied and distributed globally over the Internet primarily through P2P software. **Digital music has become ambient by bypassing all of the older manufacturing and distribution channels.**

Because perfect digital copies of songs can be made at no cost and distributed free in unlimited quantities anywhere in the world almost instantly, traditional methods and channels of production and distribution have been severely challenged. The entertainment industry has lost much of the control that made its business model successful. According to the RIAA, this has caused major financial damage to the music industry. Their response has been an onslaught of litigation against allegedly illegal file swappers (see the section "Suing John Doe" later in this Chapter.) Allegations of music piracy on a global scale have given the phrase "jailhouse rock" a whole new meaning.

Heard in the Dorm

Not everybody agrees, not even everybody in the music business, that downloads are causing the decline in music revenues. Not everybody thinks that music downloads are bad for the artistry of the music business. And a lot of people maintain that downloads of the sort facilitated by P2P software are not a violation of copyright law, despite what the RIAA says (this is a complex legal problem that needs to be settled in the courts, preferably by somebody else.) What is behind these ideas? To answer this question, we need to look at the world of music from a place where serious music fans can be found, a college dorm.

Since the introduction of CD's as a commercially viable music medium, music fans have objected to the way album content is selected for an artist. This complaint actually goes back to the days of vinyl record "albums" and "singles." A typical album or CD contains a dozen or so tracks (songs) only a small number of which are "hits" or even interesting enough artistically (except to hard-core, fanatical followers of the band) to merit more than one or two plays. This filler material is expensive for the consumer; a typical CD costs about $15.00 (about a dollar per song) and the price has gone up steadily over the past several years. In the days of vinyl, any "hits" on an album were released as "singles" which had a hit on one side and a filler song on the other but which retailed at about $.99. "Single" CD's have never been popular due in part to the high price, $2.99 to $4.99 each, more than vinyl singles used to cost even if an allowance is made for inflation. The irritation factor escalates if we look at "greatest hits" CD's, which rarely contain <u>all</u> of an artist's greatest hits. Instead, several "hits" are accompanied by filler tracks that nobody wants to hear. To get <u>all</u> the artist's good material you have to purchase another "greatest hits" CD or maybe even the "boxed set" which for some reason usually comes out just before the Christmas season and can sell for as much as $99.00. All this money spent when one CD could easily hold all the hits most artists typically produce during an entire career So, not all experts agree that declining revenues over the last half-decade are the result only of music piracy. They think music fans, the customers, are reacting to price gouging and simply spend less money on CD's.

Instead of debating the qualitative deficiencies of filler material (which are subjective anyway,) let's look at customer rejection of the product as a problem not of lack of artistic talent but rather of packaging. The music industry is not the only one under consumer scrutiny for the way they package their product. I recently needed to purchase a replacement windshield wiper blade for the passenger side of my 1997 Dodge Ram pickup. I went to Autozone. After wading through some tricky cross-reference charts (these should have been online so I could do that at home) I found the item I needed. But upon closer examination I noticed that there were <u>two</u> wiper blades in the carton, not one. Were they a matched pair, one for the passenger side and one for the driver's side? No, both were for the passenger side. Considering that the original wiper blade that came with the vehicle had lasted seven years, I was being asked to purchase a fourteen-year supply of passenger side wiper blades. I thought it was absurd and said so. The store manager explained that, in his opinion, packaging costs were the driving factor behind the wiper blade scam. A single blade would cost about as much as two blades because of packaging and Autozone felt it would blunt criticism if they at least put another blade in the package. This didn't sound reasonable to me but I went ahead and purchased the blades anyway because both blades together cost $2.97. At that price level it was no big deal.

At the price of a CD these days, music packaging scams (putting two "good" songs in with ten "bad" one) have become a big deal, especially to high school and college students who feel financially constrained. Does this justify piracy (if indeed that is what it is?)? No. Does it explain their attitude toward the music industry? Yes. The kids in the dorm (and on the street) think they are being ripped off.

Any time a business creates a marketing strategy that results in millions of customers, perhaps the majority of their entire customer base, thinking they are being shortchanged, something is wrong with the strategy. If this continues uncorrected, customer reaction will be increasing reluctance to purchase the product combined with a search for a substitute, or both. That pretty much describes the current state of affairs in music.

Another issue that bothers music fans is the fact that new bands often struggle to get the support of the major record labels. They may only have a couple of songs that are good enough for airplay but not enough to fill a CD. The new bands also drain promotional and other resources away from established acts, which are the big moneymakers for the record companies. In response, many new bands and solo artists have turned to the Internet to get exposure for their work, offering free downloads using — no surprise here — inexpensive, simple P2P technology. They encourage fans to download and sample various songs and mixes, in essence getting focus group feedback on their artistry. It works. More new bands than ever are finding an audience (students like to support new bands in part because they see them as the "underdogs" in the music business, young performers trying to climb the ladder against tremendous odds) and talent is being discovered that would otherwise remain hidden. Many established artists, taking note of the tremendous promotional power of the Internet, have started to release free P2P downloads of song fragments or even entire songs from new albums just to create fan awareness for their new material. Their hope is that they will ultimately sell more CD's and broaden their fan base. A late 2004 survey of musicians by the nonprofit Pew Internet and American Life Project indicates widespread and growing support among musicians for P2P downloading of their songs. While 47% recognized that P2P downloads might reduce their music royalties on any given creative piece, 43% felt that downloading by fans helped promote their work and expand their market. Ninety-seven (97) percent of musicians surveyed felt that the Internet did not threaten their overall ability to protect their work.

The point is that students get mixed signals. The music industry plies them with free P2P download but sues them for participating in free P2P download "piracy." The confusion within the music industry as to how to play this new technology reaps what it sews.

Turning to the music industry's business model, there have been some interesting experiments conducted in an effort to develop an alternative pricing strategy for music. Apple's iTunes Store, which sells single downloads for $.99 is the best-known example. Wal-Mart, Microsoft, Yahoo and others have all entered the business. In a sense this is a revival of the old vinyl "single" pricing strategy popular twenty years ago. Approximately 140 *million* single tracks were "legally" downloaded through these services in 2004 vs. 20 million in 2003, a 600% increase in one year.

But there is more than just $.99 downloads shaping the musical world in the digital age. Market tests have been conducted with significant price reductions on albums (CD's) and the results are interesting. For example, a CD that was being sold for the already comparatively low price of $11.00 was cut to $2.00 as an album download (again, for a limited time) and, with the new price promoted via an email blitz, sales of that particular album (as a digital file, not a CD) increased *60 times* over the prior rate! With no manufacturing, shipping or handling costs, industry experts say there is far more profit in selling 60 files at $2.00 each and raking in $120.00 than there is in selling one CD for $11.00. Rhapsody, an online music download service (www.rhapsody.com) tried a similar test with single downloads. They cut the $.99 price in half but saw the purchase volume triple. You don't have to be a college student to figure that out and modify your business plan accordingly.

So, there is hope for an amicable resolution to the student/ music industry conflict, but as far as the fans in the dorm are concerned, the burden falls on the music industry to develop a business model that doesn't drive its customers away or, worse, drive them to a life of crime.

Peer-to-Peer Networks

Peer-to-peer (P2P) file transfer software is a major breakthrough in online information distribution, especially in situations where the broadest possible distribution is desired and restrictions on data access or data security are not a primary concern. It was developed in the 1980's and brought to the attention of the computing public by the infamous Napster music web site in 1999. With P2P, an individual can publish or distribute information to or acquire information from literally millions of other computers connected to the Internet. The types of data that can be shared include self-authored content, photographs, books, articles, software, art work, spreadsheets, construction documents, music, independent films, financial reports, instant messaging, web logs (blogs), web sites and web site links. Promising business applications for P2P include streamlining business-to-business (B2B) communications and enabling collaboration on design and construction documents. Intel is using P2P to make efficient use of unused hard drive space common to corporate client PCs, which are underutilized resources in client/server distributed computing architectures. The US Government is using P2P to gather, analyze and distribute statistical data for the Consumer Price Index. EBay uses P2P to gather and distribute financial data provided by bidders and merchants. Earthlink, a leading Internet Service Provider (ISP,) is introducing P2P capability to its customers, citing growing consumer demand for P2P and the need to provide the services customers expect in order to remain competitive. Skype Technologies offers free voice calls over the Internet using P2P distributed technology. Skype, started by two entrepreneurs who created music download service KaZaA, now has 70,000 users *per* day downloading their P2P voice software and has signed up 12 million users in the past 14 months. P2P is even generating new business in support applications like Norton Utilities, which now has anti-virus software for scanning shared files over P2P networks (in the early days of P2P viruses were a major problem.) And, in what is potentially the most exciting application based on P2P, search engines are exploring the uses of P2P as an aid in searching for information contained not just in websites but also in the hard drives of PCs connected to the global online network. This could expand search results exponentially.

The low cost, extreme simplicity and efficiency of P2P make it a leading candidate to revolutionize Internet data transfer in the not-too-distant future. The number of P2P users is already very large. Edonkey, the most popular online file-sharing software provider, averaged 2.54 *million* users per day during the month of September 2004.

How does P2P work?

P2P is a network application that runs on PCs and enables an individual to create an area (called a Shared Folder) on their hard drive that can be accessed by anyone else on the Internet. The other PC drives and folders remain hidden and cannot be accessed by outsiders. Files placed in the Shared Folder can be "seen" by others, providing they have similar P2P software on their own PC. The files in the Shared Folder are available for download by any other person, without restriction, through simple drag-and-drop interface. In earlier generations of P2P architecture, a central index of available files was maintained on a web site (Napster was the prototype.) This enabled an individual user to search for files available from the Shared Folders of all PCs in the network. Newer-generation software (such as Gnutella or Fasttrack) completely bypasses a central server and enables users to conduct index searches directly on other PCs that are currently online and that have compatible software installed. This is a significant expansion of online search capability limited only by the fact that not all PCs have compatible software installed (initially this was also the case with email clients, which now come installed as part of a basic operating system. The same may eventually occur with P2P client software.) P2P download speeds have also increased, thanks not only to the steady spread of broadband but also to innovations in the software. A P2P program called BitTorrent allows full-length movie downloads in about two hours, as compared with 12 hours using eDonkey or KaZaA. Not surprisingly, BitTorrent now accounts for over half of all downloads of music and video and for more than a *one-third* of all traffic on the Internet at the end of 2004, according to the research company Cachelogic.

Whether through centralized servers and indexes or decentralized architecture, an individual can search a network that might comprise *millions* of PCs at any given moment, locate the file that they want, and then download directly from the remote PC to their local system. With the newer software, they can connect with the other PC directly and eliminate any contact whatsoever with a central point of management. There are no dedicated servers and data is not stored at concentrated points in the network. This is far more efficient (but not as secure, at least not yet) as point-to-point file transfer such as FTP (File Transfer Protocol.) Due to the large size of many files, compression software is usually used to speed up the process of file downloading. Filing sharing software is sometimes sold for a nominal fee (eDonkey is $19.95 for the complete package) but more usually given away (Grokster does not charge for its basic software, which is supported by advertising and the distribution of Adware.) Low-or no-cost combined with simple operation make P2P popular with a growing number of individual users, a trend that is expected to continue.

Should This Technology Be Controlled?

Unfortunately, the entertainment industry has launched an all-out attack on P2P file sharing software, which it believes is responsible for a significant increase in illegal distribution of music in digital file format, primarily MP3. The industry, represented by the Recording Industry Association of America (RIAA,) asserts that this sharing violates the copyrights of recording artists. There is a lot of money at stake. Recording industry revenue losses tabulated by the RIAA are running at about 14% of total potential annual recording industry revenue, approximately two *billion* dollars per year.

The most highly publicized action undertaken by RIAA to suppress P2P has been a series of lawsuits against web sites and individuals (including students) by the RIAA and its member companies. The best-known of these was the first one, a lawsuit against Napster, a web site that in 1999 started promoting the use of their P2P software to facilitate file transfer between and among individually-owned PCs. The RIAA sued in 2001 and won when a court shut down the Napster site. Napster, which defended itself mostly in the public forum using what can only be called a "media circus" strategy, was found guilty of contributory infringement because it maintained online indexes of files available from its members. The court ruled Napster *could* have taken steps to curb illegal swapping but did not. (Napster was subsequently bought out by another company and has re-emerged as a "legal" P2P software site.)

Napster turned out to be the only "easy" target for the RIAA. Subsequent lawsuits against better organized and more professionally defended P2P software companies such as Grokster (whose web page motto is "Support the Artist. Buy the Record.") have failed. The Grokster victory was reaffirmed by the United States Court of Appeals for the Ninth Circuit in August of 2004. The gist of the ruling was that the mere fact that a device or a technology is capable of being used for illegal activities is not sufficient reason to suppress the device or technology. The plaintiff (actually it was RIAA member MGM that had brought the lawsuit, not RIAA itself) had argued that, since because the vast majority of users of Grokster file sharing software were conducting file transfers that (according to RIAA) violated copyright laws, the software was willfully developed and distributed free of charge to promote these allegedly illegal activities. The judges disagreed, harkening back to a well-known suit filed against Sony by Hollywood studios (*Sony Corp. v. Universal City Studios, Inc.*) charging that the sale of videocassette recorders constituted contributory copyright liability, since buyers could use them to tape movies and other proprietary content. The Supreme Court ruled in 1984, in what has become known as the "Betamax decision," that because VCRs were capable of providing "substantial non-infringing uses," Sony should not be held liable merely because some customers might use the product to copy things they shouldn't.

In the opinion of the judges in the Grokster case;

"Further, as we have observed, we live in a quicksilver technological environment with courts ill-suited to fix the flow of Internet innovation. The introduction of new technology is always disruptive to old markets, and particularly to those copyright owners whose works are sold through well-established distribution mechanisms. Yet, history has shown that time and market forces often provide equilibrium in balancing interests, whether the new technology be a player piano, a copier, a tape recorder, a video recorder, a personal computer, a karaoke machine, or an MP3 player."

Suing John Doe

Having failed to close down web sites such as Grokster and cutting off P2P at its source, the RIAA has taken the unusual step of filing lawsuits against individuals, including many high school and college students, whom they accuse of sharing copyrighted music files with P2P software or by buying, selling or trading "bootlegged" CDs at "swap-meets". This is a risky strategy, not because the RIAA expects to lose any of the lawsuits (although they could) but because it has angered the music buying public, the very people the RIAA hopes will purchase more CD's and music videos. It has also angered some of the recording artists themselves, who oppose litigation against their fans and who are happy to make digital song files available for download via P2P protocols. Many RIAA artists now encourage the very same activities that are cited by RIAA as justification for suing their fans.

The RIAA tactics in identifying potential defendants in these lawsuits has been to use P2P software to search through the Shared Folders on thousands of individual PCs. Since these folders are, by definition, public domain, this appears to be perfectly legal. When a Folder is identified as containing copyrighted materials that appear to have been obtained by sharing, the owner of the PC hosting that folder becomes a target for RIAA. They either receive a warning that they will be sued unless they agree in writing to cease and desist and pay damages, or they become a defendant in a lawsuit. Unfortunately for the RIAA, identifying the IP address of a specific PC does not necessarily mean that they know the name or physical address of the owner, so many of the lawsuits are filed against "John Doe." To address the matter of identifying alleged infringers, RIAA sued Internet Service Providers (ISP's) such as Verizon to force them to turn over the actual name and physical address of the "John Doe" targets (who are customers of the ISP) prior to the lawsuit being filed. In April of 2003 the RIAA won a victory when a court ordered Verizon to divulge the names of some alleged infringers. Verizon appealed, however, and in December of 2003 a federal court overturned the initial lower court ruling, saying that Internet Service Providers cannot be compelled to disclose the identities of customers suspected of illegally sharing copyrighted songs. While acknowledging that Internet "piracy" is a crime and a significant problem for the entertainment industry, the court nonetheless said that Verizon couldn't be held responsible for material that passes through its Internet network. The judicial opinion said that the RIAA's argument that ISPs could be subpoenaed because of the material its customers transmit on the Internet "borders on the silly."

RIAA has not given up. They continue to litigate against individual John Does thought to be infringing on copyrights. They have been filing a lawsuit at the rate of one per month, each one with several hundred alleged violators. On Sept 30, 2004, for example, a lawsuit was filed against 762 persons including individuals at 26 different universities across the country. Having lost the Verizon case, RIAA has been forced to use the John Doe litigation procedure, which is used to sue defendants whose names are not known. There were also 68 named defendants. According to the RIAA press release that accompanied the lawsuit "32 individuals at 26 different schools were sued by the major record companies for using their university networks to illegally distribute copyrighted sound recordings on unauthorized peer-to-peer services. During the past four years, illegal file sharing, particularly on college campuses, has been rampant" The individuals included in the Sept., 2004 legal action were on the networks of the following universities: Appalachian State University, Augsburg College, Claremont McKenna College, Colgate University, College on Mount Saint Vincent, Columbia University, Georgetown University, Hampton University, Illinois Institute of Technology, Kean College, Kent State, Louisiana State University, Michigan State University, Minnesota State University, New York University, Pacific Lutheran University, Portland State University, St. John's University, Stanford University, State University of West Georgia, SUNY College at Old Westbury, University of Connecticut, University of Louisville, University of the South, Virginia State University, and Western Illinois University

Although many legal experts agree that unlimited and indiscriminate sharing by individuals of copyrighted digital files may violate Federal copyright law, nevertheless, the merits of the allegations are unknown because defendants do not have the wherewithal to have their day in court against the almost unlimited resources of the entertainment industry. They plead no contest or guilty or accept an "amnesty" offer from RIAA and pay a damage claim. The primary rationale for these lawsuits has not been restitution for damages. It has been *publicity*. The RIAA hopes that intimidation will reduce the number of allegedly illegal downloads from digital file sharing web sites. They also hope it will reduce the extent to which Internet users swap files directly among themselves using P2P software. Cary Sherman, the RIAA's president, has said the lawsuits against university network users in particular are designed to drive home the message to students that unauthorized downloading has consequences and that legitimate alternatives are available. RIAA cites evidence that this strategy is working. A July, 2004 survey by Peter D. Hart Research Associates found that the percentage of those polled who know it is illegal to "make music from the computer available for others to download for free over the Internet" is 64% illegal to 13% legal. The July Hart data also revealed that, by a margin of 60-17, those polled are "supportive and understanding" of the tactic of legal action against individual illegal file sharers; a 55-17 margin believe there are "good legal alternatives"; and only 31 percent think the activity should be "legal," compared to 56 percent who say it should be "illegal." The RIAA sites a 4.5% decrease in P2P downloading in the first half of 2004 and a 3.6% increase in digital music sales during the same period, which they say are related. Market researcher Soundscan reports a higher increase in album sales — 5.9% — during the same period. However, other surveys dispute the notion that downloads have decreased since the wave of litigation began. A consulting firm in Atlanta — BigChampagne — shows a 13.2% increase in file sharing during the first nine months of 2004 and indicates that November of 2004 was the biggest month for P2P activity in history with an average of 5.5 million sharers online at any one time, up from 2.5 million during the same month, prior year. BigChampagne credits the recent upturn in album sales to the increased exposure music is getting through P2P downloads. It is hard to say what numbers are valid. The Hart survey was commissioned by RIAA; BigChampagne's business model depends upon extensive P2P sharing. But RIAA insists they are on the right track.

The Motion Picture Association of America (MPAA) agrees with the overall RIAA strategy and is taking it several steps further. Starting in December of 2004 they began suing "individuals who set up and operate computer serves and web sites that, by design, allow people to infringe copyrighted motion pictures." The primary target is web sites that use BitTorrent technology. "The target of our actions is not technology," John Malcolm (head of the Motion Picture Association of America's antipiracy unit) said. "There are many legal Torrent sites…that are dedicated to the distribution of public domain work and we are taking no action against them whatsoever." Some of the web sites named in the lawsuit are in Europe and are not subject to U.S. copyright law. MPAA is also seeking injunctive relief from the courts against two file sharing web sites, Grokster and Morpheus. Lower courts have repeatedly refused to shut these (and similar sites) down so MPAA petitioned the U.S. Supreme Court to hear their case. In December of 2004 the Supreme Court accepted the case, setting the stage for what MPAA (and RIAA) hope would be a landmark ruling. A ruling is not expected until sometime in late 2005. It should be noted that MPAA fought vigorously against VCR's in the 1980's, declaring that if courts allowed the technology it would devastate the movie industry. In the 1984 Betamax decision, the Supreme Court was unsympathetic to this argument, which proved to be wildly inaccurate. The new Supreme Court case (*Metro-Goldwyn-Mayer Studios Inc. v. Grokster, Ltd.*) will be, in essence, a revisiting of the Betamax case.

Is the copyright issue on its way to amicable (albeit fear-based) resolution and can we all relax? No. First RIAA is pressing on with renewed efforts to get legislation passed that would reverse the Verizon verdict. They have promoted legislation called the *Inducing Infringement of Copyrights Act* (S.2560) which would hold liable for copyright infringement any company or person who "aids, abets, or induces" the illicit sharing of copyrighted works. This would overturn the Betamax decision of 1984. RIAA has also teamed up with the MPAA to get the Federal Communications Commission (FCC) to mandate that copy-protection technology must be built into some home entertainment devices such as TIVO and DVD recorders. To assure that companies comply, the FCC requires them to show MPAA new technology being developed and gain approval from MPAA before the devices are released for sale to the general public. This has set the stage for new legal battles, pitting MPAA (and RIAA) against companies like Microsoft, Apple and RealNetworks, who claim the FCC regulations will cripple innovative technology through intimidation and the loss of their right to keep innovative software confidential until it is released. Academic computer experts and private security consultants have also been targeted for publishing papers that explore computer security issues and sometimes expose problems with MPAA/RIAA-approved anti-piracy software being used by hardware and software developers. MPAA claims that by making this information public, hackers and other violators gain valuable information that helps them defeat security features that benefit MPAA and RIAA. A lawsuit and counter-lawsuit between RIAA and a Professor at Princeton University in 2001 ended in a standoff, but the issue remains unresolved.

A Vision of Things to Come

While RIAA litigates, the probable future of P2P comes into clearer focus when we learn that many RIAA member companies are already quietly finding innovative ways to integrate P2P technology into their business model. In Atlanta, BigChampagne monitors P2P music file transfers for RIAA music companies, not to catch alleged copyright violators, but to find out which songs and artists are most popular as indicated by the download activity occurring. They monitor the millions of downloads that occur (BigChampagne estimates that there were approximately 8.2 million P2P downloaders online at any given moment in July of 2004, up 25% from the prior year) and cross-reference downloader IP addresses with zip codes to find out what is popular in any given media market. Music companies pass this information on to radio DJ's and encourage them to adjust their play lists to better reflect local listener preferences. This boosts their listener ratings. Not coincidentally, this also boosts CD and concert ticket sales for bands. New bands are especially appreciative of this kind of marketing research without which they might not get the radio exposure needed to establish their commercial credibility.

In essence, P2P downloaders now serve as a focus group that provides insights into pop music trends from a perspective not previously available to record company executives. But there is a downside as far as the RIAA is concerned; the record industry's lawsuits against file-sharing companies are based in part on their assertion that P2P software (unlike many other copy devices) has no use other than to help infringe copyrights. If the record labels acknowledge a legitimate use for P2P programs, it could undercut their case. Perhaps for this reason, the labels are not very open about their use of P2P market research, but companies who are alleged to be using these types of services include Atlantic, Warner Brothers, Interscope, DreamWorks, Maverick, Elektra, and Disney's Hollywood label, major players all. Conclusion: RIAA faces major revolt from within if the users of P2P marketing research appear to gain advantage in the sale of CDs and the promotion of bands over companies who do not use the technology.

"Legalizing" P2P

There has been widespread collateral damage from the copyright wars being waged by the entertainment industry against file sharing companies, their technology and the individuals who use them. Individuals and organizations are intimidated. They may also have been convinced that the only potential uses of P2P are in entertainment applications, whether audio or video, legal or illegal. This is not the case. While legal issues should not to be ignored, there is ample evidence even from the legal perspective that P2P has a big future. Organizations of all types but especially universities and their library systems should make certain that they preserve their options re: P2P and are in a position to capitalize on the future uses of this technology which is already impacting digital processes ranging from search engines to archival data distribution. The challenge is to not get caught in the meantime in legal battles that have yet to subside.

An attorney will tell you that the best legal strategy is a good defense to neutralize the risks of litigation. Several IT and administrative options exist, including the following:

1. Put all network users on notice that requests from RIAA or their affiliates or other *bona fide* copyright holders for IP addresses of individual computers running on, for example, a university network, will be made available. Not only will this make it clear that those accused need to consider their actions and defenses beforehand but also that any expansion of P2P activities on campus is not to be misinterpreted as approval of unrestricted downloads of copyrighted materials. This is not an acknowledgment that downloads violate copyright law. But until this matter is clarified by some court cases, the network is not a safe haven for possible illegal activity.

2. Block access to P2P web sites. This has been criticized as an anti-democratic strategy, tantamount to censorship. In the Internet world, this is frowned upon. Google, for example, was severely criticized for cooperating with censorship in China. It might not withstand a legal challenge and is sure to be unpopular, but it would certainly meet any test of sincerity about short-circuiting certain types of P2P networks.

3. Bandwidth — the amount of data that can be transmitted over a network at any given time — can be limited for certain types of file formats, specifically MP3 (music) files. By slowing down data transmission, access is preserved for other types of files but for all practical purposes the appeal of music downloading is lost. A single MP3 can be slowed down so that several hours are required to download. Meanwhile other file formats can proceed at normal speeds, so that overall information transfer through the network is not degraded.

4. "Packet shapers" can be employed to slow downloading and uploading at specific web sites or for specific IP addresses. Again, a single MP3 can be slowed down dramatically.

5. *Legitimate access to music downloads can be provided as an alternative to piracy.* This would cost a little money but would make everybody happy. This is the approach best calculated to deflect legal action. It is heartily endorsed by the RIAA and they are signaling strong approval of legitimate download arrangements between, for example, universities and music companies. "There have been many exciting developments on the university front in recent months," says Cary Sherman, the RIAA's president, as quoted on the official RIAA web site. "An ever-expanding number of school administrators, often at the behest of students, are signing partner-

ships with legitimate online music services. Students get the benefit of high-quality, legal music while schools get to spend less time worrying about their students getting into trouble. It is a win-win for everyone." Compared to the beginning of last year's (2003) academic year, when there were no such agreements, there are now (starting the 2004-2005 school year) more than twenty universities that have established a partnership with what the RIAA considers a legal music provider. Ironically, one of the legal download sites that has been most aggressive in forging relationships with colleges and universities is Napster, the original target (and sole victory) of the RIAA. Napster offers a low-cost listen-only subscription service with a $.99 download option. The listen-only service keeps the user "tethered" to the music files by streaming them for playback but not allowing files to be saved to a local hard drive. Listen-only use is unlimited. Universities with arrangements with Napster at the beginning of the 2004 academic year include Cornell, University of Miami, Vanderbilt, and Penn State. Ruckus Network, a Napster competitor that offers movies as well as music for a monthly flat rate, has signed up Northern Illinois University, Alfred State College and Bentley College. Such arrangements do not protect all students at these institutions from litigation, but the institutions themselves would seem to be protected and are therefore free to make use other P2P applications for information collection and distribution.

As important as music is to people, nevertheless, P2P is not just about music (as RIAA would have us believe.) It is also about search and distribute (publish) two functions that make the Internet a platform without equal for academic information services. The danger is that fear of litigation will slow the adaptation of P2P (and wireless networks) by legitimate institutions like colleges and libraries that, after all, are in the information business. Appropriate strategies that minimize the risk of legal entanglements need to be selected and deployed. P2P, connected wirelessly, should then be encouraged to become what it is becoming anyway — a mainstream application for the broadest possible array of digital communications tasks.

7. Digital Students

Today's students are prodigious consumers of digital entertainment and information and accomplished digital communicators in ways that most administrators, Faculty and parents are not, at least not yet. They arrive this way from their high schools and bond digitally with their peers on campus to form yet another distinct class of "digerati" (digital communicators.) Even the social scene for students is increasingly based on digital interaction. Their familiarity with digital communications techniques has a major effect on their academic expectations and achievements as well. Digital students are increasingly at odds with the analog world, which often seems too unresponsive to keep up with the digital pace of their daily lives. Conversely, it is becoming more difficult for an analog world to communicate with students, who seem to relate in a different way to information resources. An appropriate response designed to avert a complete disconnect between the digital and analog worlds is critical. But what are student expectations when they arrive on campus? Who is determining their academic and social information needs and what is the best way to do this?

Successful marketing strategies are developed with the assumption that customer needs and expectations must be identified before they can be fulfilled. In the private sector, market researchers constantly monitor customers. This is mostly done with surveys but also may include direct observation of customer habits (how they walk through a store, for example, or what products are purchased in conjunction with other products.) The Internet has made surveys easy and inexpensive and web surfers are regularly bombarded with requests for feedback. Many web sites contain blog-like areas where customers can write "reviews" and "evaluations" of various products or services. These survey-bloggers read each other's comments and can then thread comments on comments, taking the survey far beyond a simple question and answer format. This can evolve into a mother lode of information for the company conducting the survey. More traditionally, customers are invited to join advisory boards or to "vote" on various products and services. Advertising takes up the cause and turns the "survey" into a "sales pitch" ("great taste...less filling," which is it? They want you to think it is both.) Customers are flattered to think that their thoughts and opinions are important to a company and surveys about surveys show that customers think more highly of a company that solicits their opinions even if none of their ideas are ever used. Whole ad campaigns can be built around a seemingly open-minded request for customer feedback even when the actual responses have little or no real value.

On the other hand, customers are smart and often have goods ideas and instincts about the value of products and services, especially when their responses are considered in the aggregate sense. Marketing experts who have been in business and have surveyed customers will tell you that any given idea suggested by a customer is almost always bad (impractical, expensive, lacking advantage or benefits to the company or to a large group of customers, etc.) Many are unbelievably silly. But, it would be a mistake to assume that, just because any one idea is bad, then all of them combined must be *really* bad. This is not the case. Collectively, customers can often point to emerging trends or needs that the marketing manager might otherwise never see. Customers are also an excellent source of competitive and comparative information, either about other companies that are offering similar products or services or, more importantly, alternative ways of accomplishing the same thing.

In education, the administration and Faculty have traditionally determined student needs with some outside assistance from parents. But if administration, Faculty and parents are not attuned in to the students' digital world, a disconnect can occur. And, considering that this digital world is rapidly changing, knowing what students think is an ongoing challenge. Are the students themselves studied through "consumer marketing research" and does anybody making decisions about their future know what *they* expect or think they need when it comes to classes or career preparation? Are the opportunities in the job market or graduate school placement taken into consideration? Are the digital communications preferences of each new class of students studied? What new ideas does each new class of students bring to the campus? If these questions are not being asked and the results analyzed, a university can fail for the same reasons a business can fail…by losing touch with its "customers."

Let's look at the new world of the digital student through the eyes of marketing researchers

©Reprinted with special permission of King Features Syndicate.
Analog parents-digital child.

<u>Understanding Millennials</u>

How "digital" are students when they arrive on campus? Students start out ignorant of much of the academic material they will be exposed to while they are at college but they are already experienced digital communication experts. Today's student generation is known to marketers as "millennials," or, when lumped together with young professionals in their 20's, the Thumb Generation (because many of their digital devices are held with both hands and operated with their thumbs on the keyboard.) They have grown up with the Internet and cell phones as household appliances, so they are far more information savvy than prior generations. They expect, even demand that information transfer occur digitally. They also tend to group all information—academic, political, social, entertainment, etc.—together, and look to the same sources for all three (late night comedy shows have become a major source of political "news," for example.)

It is second nature for most millennials to multi-task their way through information sources throughout the day, combining music, course content, news, sports, games, and personal communications via audio, video and data formats using cell phones, TV, TIVO, iPods, DVD and CD players, Xboxes, and maybe even PDA's (Personal Digital Assistants, which are used more by young professionals than students.) According to recent (2004) market surveys, 80% of all students carry cell phones and 36% use them for text messaging (called Short Message

Service—SMS for short) as well as voice communications. The volume of communications activity by cell phone is mind-boggling; there were 23 *billion* SMS messages sent via cell phones in 2004. (An example of a text message might be T+&UBOK—"think positive and you will be OK.") Millennials like to customize their cellular ring tones with 10-second snippets of popular tunes, not just because it's cool but also so they can tell, when cell phones are ringing all around them, if it is their phone. Called "ring tunes," different ones can be assigned to individual callers so they also know who is calling, sort of an audible Caller ID. "Ring backs" are song snippets played back to the caller from the phone of the person being called. At about $2.50-$3.00 each, the custom ring tones and ring backs and related services will exceed $3.8 *billion* in sales by 2008. Cell phone "wall paper" (the background that appears on the color screen of a cell phone) is also becoming popular. Ring tunes can be matched with wall paper, so that the cell phone user has a complete, customized, audio-visual experience each time somebody calls.

About 80% of all students connect to the Internet via broadband, putting multimedia downloads on a par with their use of TV and radio (the pre-millennial generation was overwhelmingly dependant upon TV for electronic information. For earlier generations, radio was the dominant electronic information source, available everywhere once the transistor was developed.) Average online time is three (3) hours per day as compared with only one hour and 42 minutes watching TV. Of the 3 hours, 57% of the time is used for email activities and 43 % for web browser activities. A big part of browser time, about 20%, is spent playing online games. Ninety-nine (99) % of today's millennials use email, an estimated 50% blog regularly, 59% use instant messaging and chat rooms, often while they study. (Habits formed in school carry over into their professional lives as well. An October 2004 survey of employees shows that 25% use instant messaging for personal communications during a typical workday.) Fifty percent of all 13-24 years-olds simultaneously watch TV and surf the Internet when taking a break from studies. PVR's (personal video recorders) are projected to be the next big dorm room appliance for incoming college classes in 2005 and 2006.

A study of 4839 Internet users conducted by research firm Knowledge Networks at the end of 2004 indicated that, on average, 14 minutes per day was spent "dealing with computer problems." This adds up to about 10 8-hour days per year. Suppose you did mechanical work on your auto for 15 minutes every day or studied Italian for that amount of time each day all year long. After awhile you would probably know a great deal about auto mechanics or Italian. It is little wonder why today's millennials are so knowledgeable about and adaptive to software and hardware. They not only use it for hours each day, they work on it every day to keep it operating.

The rate of change in information handling expectations has become so rapid that campus juniors and seniors may find themselves behind the incoming freshmen when it comes to "techo-savy." For example, 21 year-olds accept being connected to the Internet via "land lines" (hard-wired, physical connections) whereas 16 year-olds are more likely to think that wireless (Wi-Fi) connectivity is the standard. In a couple of years students will expect not just broadband and not just wireless but wireless "very broadband" (called Wi-Max) with speeds much greater than those currently available. A peculiar version of Moore's Law (which in 1965 postulated that the number of transistors that could be placed on a CPU would double every 18 months and therefore the speed of computers would double at the same rate) seems to apply to student digital culture at large; the complexity doubles every eighteen months. Universities will have at least two and perhaps even three "generations" of millennials on campus during any given four year period, each with distinct technological habits and expectations.

The digital pace of millennials is becoming so intense, so frantic, the ever-expanding supply of digital gadgets so overwhelming, that there are actually some signs of a backlash. Millennials are starting to have an occasional cellphone-free party so they can actually talk to each other. For commuters, Amtrak has started putting a "quiet car" on some of its trains where no audible digital communications or cell phones are allowed. There are even rumors that the use of the "vibrate" ring mode is gaining popularity as a defense against jangled nerves caused by incessant electronic sounds.

The challenge is to communicate with each of these generations on their terms as well as those traditionally favored by the university. Instructors cannot expect to teach using nothing more than a Xeroxed syllabus, some reserved books at the library and a semester's worth of stand-up lectures. Nor should libraries expect students to be enthusiastic about physically take books from the shelves, pass through a "check out" process, use the books at a remote location and then physically return them to the library building. Today's "millennial" students simply do not acquire information that way.

A Digital Career

The career prospects for today's millennials are changing almost as fast as the technology that they must master. It is a gross understatement to say that the economy is trending rapidly toward an information-capable workforce. Some examples to illustrate; online, real-time collaboration is becoming the standard for design, construction, advertising and virtually all forms of project development in most businesses. Intranet publishing of documents by authors is standard procedure at legal and consulting firms. Internet publishing is the new standard for Faculty as well as all of us liberal arts graduates who like to write, draw, photograph, and play with ideas in a creative manner. (The book you are reading is being published and then edited online. It resides on a server in Virginia, not on my local PC. It may or may not ever be printed on paper or put on a library shelf, yet it will be instantaneously available to tens of millions of people simultaneously, if that many are interested.) As MBA and business majors are discovering, all types and levels of business management are becoming impossible without a complete grasp of digital communications practices. And, of course, all 21st century technical fields, from the physics of nanotechnology to the agricultural sciences, are a complex blend of empirical studies and digital information processing.

This goes beyond what used to be called "computer literacy." All of these careers require more than just a mastery of certain software packages or hardware configurations; they require a completely different mindset as to how information is gathered, stored, analyzed, processed, protected and distributed. The impact of these career requirements on job prospects for millennials and the thumb generation is enormous. They cannot afford to purchase an educational "product" that short-changes them by graduating them as digitally inexperienced or unaware. In the age of total information traditional educators who show little concern for the post-college needs of their students provide a serious disservice. Pity the poor 21st century soul wandering around campus with a tattered Chaucer anthology expecting to find a real calling or earn an adequate living after graduation. This student's life prospects would be far better served if he/she at least knew how to download Chaucer in .pdf format onto a wireless tablet.

Testing for Digital Literacy

The ability of students to demonstrate adequate digital literacy is becoming a factor both in college admissions and job search after college. The non-profit Educational Testing Service (ETS,) the people who provide colleges with the Scholastic Aptitude Test (SAT,) Graduate Record Examination (GRE) and other admissions tests, has developed a new test called *The Information and Communications Technology Literacy Assessment Test* that attempts to assess students' ability to sort, assess, manipulate and ultimately make good critical evaluations of an array of information of the sort available to them from the Internet. This test will be given (in beta format) to about 10,000 students across the U.S. in the early part of 2005. ETS says both colleges and businesses have requested a test of some sort to help them evaluate admissions and job applicants because of concerns about a decline in critical thinking skills. In 2001 ETS, in consultation with educators, technologists and government representatives began a process of defining the core characteristics of information consumption at the college level and hope to eventually develop global standards in this area. Critics are skeptical about whether or not a test like this can be effective, indicating that ETS' main objective may be only to sell testing services. But some of the questions are interesting and challenging. Here is an example provided by ETS (keep in mind that this is a college entrance question.)

1. Reorder a table (provided by the test) to maximize efficiency in two tasks with incompatible requirements.

Successful college graduates can expect more of the same once they enter the job market. Google's GLAT is an example. If you haven't applied for a job lately (I have not been interviewed or filled out a job application in more than 25 years) you may be in for a surprise when you see what candidates are expected to know and show. Even "soft" jobs that do not require sophisticated technical training or experience want applicants who think logically and can gather information from a myriad of sources to attack problems with computer-like efficiency. Want a job? You could begin cruising the more than 1800 online job sites where employers from around the world post openings. Or you might try answering some of these questions from a hiring aptitude test called GLAT (Google Labs Aptitude test) used by the software company to recruit employees. While this has been panned as a "funny riff on the standardized tests that pervade academic life," Google thinks this is serious business. The test was published in magazines such as *Mensa* and the *MIT Technology Review* read by college students and young upwardly mobile professionals. Google even awards prizes and trips to their company headquarters for the most outstanding applicants. The first prize was $10,000 and a job offer (the winner of the 2004 contest took the money but turned down the job.) While some of the questions obviously require technical knowledge that liberal arts majors might not have, nevertheless, some are essay questions that should be just right for a 21st century creative mind trained in a digitally enhanced liberal arts tradition. Sharpen your pencil and let's get started.

1. Write a haiku describing possible methods for predicting search traffic seasonality.
2. This space [an empty box on the page] left intentionally blank. Please fill it with something that improves on emptiness.

3. What will be the next great improvement in search technology?
4. What is the coolest hack you have ever written?
5. What is the optimal size of a project team, above which additional members do not contribute productivity equivalent to the percentage increase in the staff size? [Options: 1, 3, 5, 11, 24]
6. What is broken with UNIX and how would you fix it?

In the interest of time and also to protect everybody's intellectual ego we will stop here. You don't have to tell anybody how you think you might do on the actual test (which is much longer) but you now have a better idea what the millennials face when they graduate to the outside world. It may be fun, even funny, but it is not easy, and it assumes a sophisticated fluency in digital communications. Google said there were lots of people who scored brilliantly on this test. The winner was chosen from 50 finalists. Google was pleased with the quality of recruits and plans to hold a contest like this every year.

8. THE LIBRARY AS A SERVICE BUSINESS

The Non-Profit Budget Crunch

A lot of attention has been given recently to the idea that a university or college (or even a high school or grade school) can be organized and managed as a for-profit business. Most traditional not-for-profit institutions face a perpetual budget crisis so it is not surprising that profitable, privately owned universities, and especially those online such as the University of Phoenix and Jones International University, have caught the attention of legislatures, taxpayers and tuition payers. The underlying principle of the for-profit universities appears to be that the product (a degree) is more important than any attributes the institution awarding it might have (on-campus life, face-to-face contact with Faculty, etc.). They sell credentials, not "atmosphere" (as on-campus experience is sometimes called by the proponents of distance learning.) The size and scope of for-profit education has grown to the point that it now challenges the very concept of public or non-profit education. The University of Phoenix, for example, had 268,000 students at the end of 2004 (133,000 of them online) and generated $336 *million* in profits on $1.8 *billion* in revenues. The stock of parent company Apollo Group, Inc. is publicly traded (Nasdaq APOL) and had a market cap of $14.68 *billion* at the end of 2004. Tuition for a full-time student is about $10,000 per year. What is most astonishing is the marketing budget used to recruit new students. The U. of Phoenix spent $383 *million* on sales and marketing in all of 2004. This amounted to 22.5% of total revenues in the 4th Quarter of 2004. This is huge even by for-profits standards. In 2004 the U.S. Education Department accused the University of encouraging a high-pressure sales culture complete with "telemarketing boiler rooms" and paying salaries and commissions based on the numbers of new students that recruiters in its Admissions Office were able to sign up. Recruiting incentives violate federal law and jeopardized the University's eligibility to participate in federal student loan programs. The University settled the case and paid $9.8 *million* in fines without admitting guilt.

But the hardball tactics of the U. of Phoenix and other for-profits is getting results and non-profits are destined to feel the pinch.

Inspired by the for-profits, the not-for-profits are looking for ways to "market" various services and create new "revenue streams" that reduce budgetary pressures. Cornell University, for example, sells reference desk librarian research services to alumni (actually, to anybody willing to pay up to $135.00 per hour for help.) University Foundations, especially at larger institutions with high-profile sports programs, "sell" the whole university through various marketing and promotional programs, mostly aimed at alumni and the families of students. There is heated debate over the wisdom of turning entire academic institutions into branded products that raise huge sums of money from alumni and the parents of students. But the money is flowing into the universities at an estimated rate of a quarter of a *trillion* dollars annually. Given the amount of money involved and the competition from for-profits, the trend is almost certainly irreversible.

Proponents of various degrees of "privatization" have suggested that State budget allocations for public institutions should be reduced but that in return the institutions should be given greater operating and governance latitude than has traditionally been the case. Currently, approximately 80% of all U.S. colleges and universities are public institutions supported primarily by taxpayers. Under proposals being debated, we would probably not see total university privatization. But a new type of mostly private but partly public institution could evolve, perhaps under some sort of "charter" arrangement so as to preserve non-profit status (a huge tax benefit to private donors as well as to the institution itself.) This hybrid institution would rely on a combination of private fund raising, competitive public and private research grants, increases in tuition and fees, and revenue streams generated from for-profit activities. These funds would be combined with annual few-strings-attached grants from the State budget. Concerns have been raised that if this were to happen, it would primarily benefit the larger, better known public universities that had strong "alumni support" (code for a successful football or basketball team) and the overall financial ability to provide for their own budgetary needs through endowments, etc. The concern is not only that the smaller colleges would be unable to compete but also that many universities, given the authority, would be tempted to raise tuition and fees substantially to create more revenue and in doing so price themselves out of the range of lower and middle income students. They would become elitist, which is a charge often leveled against current private institutions. In fact, this may already be happening. Nationally, in 2003 state and local governments now contributed only 64% of all public institution operating funds, down substantially over the past ten years. The balance comes from private sources. The University of Virginia, for example, now raises 92% of its annual $1.7 *billion* dollar budget from non-State sources, receiving only 8% from the State. Virginia is seeking reduced State controls so that it has more flexibility to, among other things, raise tuition which is currently set at $6,600.00 (versus the national average of just over $5100.00) but which UVA wants to increase to $10,200.00 over a five year period. The critics complain that UVA, already an elite school, aspires to Ivy League status with a student body to match. This would mean higher average student family incomes as well as high SAT scores at UVA.

Profit vs. Value

These are controversial issues that have mobilized advocates on all sides. Predictably, there has been collateral damage, including the emergence of a legitimate, well-articulated bias against applying "business practices," especially the ugly specter of "profits," to the management challenges of the university and its programs. As challenges grow, however, any and all strategies — including those borrowed from the private sector — need to be considered. An automatic rejection of the strategies and tactics used to create value in a private sector entity is shortsighted.

Businesses are graded by an accounting of profits, but business success is also measured in other ways, the innovative nature of products, for example, or the aggressive pursuit of customer satisfaction with services provided. As someone who has owned many smaller businesses, I can tell you that there are times when profit isn't even all that important if cash flow is good. (You have probably seen the popular bumper sticker — "Happiness Is Positive Cash Flow.") For larger companies, overall "shareholder value" is the top priority at many companies, a metric that includes profits but also takes into consideration many other things. In fact, shareholder value, a term which usually includes an allowance for the organization's "reputation" (called "goodwill" by accountants) is largely based on perceptions held by the stakeholders about the organization and its products and services (stakeholders include shareholders, non-shareholder investors such as bondholders, lending institutions, customers, employees, management, vendors, outside analysts and, of course, pundits.) There is a significant "future factor" in the perceived value of a business— where is it going, what are its goals, what is its energy level, how good is current management, what are its overall prospects? It is shortsighted for public sector institutions to reject business strategies and tactics simply because of distaste for the dynamics of profitability. There are many other business management ideas that can be borrowed without turning an institution into a profit center.

What follows is not about profits. It is both narrower (focusing on the university library and not the whole university) and shorter-term (program development and implementation over a period of two-to-three years.) What I am suggesting is that library systems can aggressively co-opt the new digital opportunities available and use them to alter and improve the <u>service</u> it delivers to Faculty and students. It can do this by looking closely at how successful (for profit) information service businesses have responded to the same opportunities. Continuing improvements in library services, even small ones, add to the quality of education delivered to students and also enhance the reputation of the university as a whole as perceived by all stakeholders. In regard to an accounting of profits *per se*, I agree with the critics that there are many differences in how a university library is administered as compared with how a business is managed, especially in how results are measured. So, I am not suggesting that library services should be "sold" or that the library should become some sort of "profit center" within the larger university. But, metrics aside, the blending of private sector thinking with strategies for improving service and managing change in a library system creates a helpful framework for aggressive program implementation. The way a successful service business operates can help shape a model for change and risk-taking in a university library environment. The analogy might also encourage new (even counterintuitive) thinking because it advocates ways of identifying and embracing opportunities for change that otherwise might not occur to a university administrator or a Faculty member, given the way they normal functions within a university. But once it is recognized that we have entered a new age of information transactions where students expect all information to be digital, universally available, individualized and (mostly) free, it is well worthwhile for the university library system to consider how the private sector organizes and delivers these services.

Let's examine this "business model" more closely. In the analogy;
- The library is the "business unit" (delivers the product to the customers.)
- The students are the "customers." (consumers of the product, a substantial amount of which is information or dependant upon information.)

The business unit "bridges the transaction" (as they say in marketing) between the customers and products and services produced or created or specified by the suppliers.

Students as "Customers"

One of the main reasons a university exists is to educate students. As recently as the mid 20th century colleges were fewer in number and the student population was smaller, consisting mostly of a socially elite group of interconnected families and their associates. Today, post-high school education is available to the masses. Not surprisingly, education now tends to be mass-marketed. Academic content and an educational environment are blended together by the university administration and Faculty to create a "product." This product is then "sold" to the students and their parents through year-round promotional and marketing campaigns. The *perceived value* of this product (as compared with the product being offered by competing universities or the alternative of skipping college and entering the military or labor force) is what initially brings students to campus and then keeps them there until they complete their degree requirements. Universities spend a fair amount of time and money promoting their product to their potential customers. It is reasonable, therefore, to start by looking at students as the university's customers. There are some obvious differences between students and customers, but a comparison will be interesting.

The conventional wisdom among university Faculty and administrators is that, when it comes to using university services, students are not like customers because they cannot react unpredictably the way customers often do. Customers can make rapid, knee-jerk responses either toward or away from products and services, reacting quickly if product quality falls off or if the products or services provided do not meet their needs or expectations. Students, on the other hand, have to be attracted to the campus initially but, once they get there, would seem to be "locked in" for an extended period of time…for a course, for a semester, for a degree, etc…and for a fixed array of courses concentrated in a few departments that are required to complete a degree.

However, upon closer examination there are some clear similarities between customers and students. First of all, private sector customers may switch grocery stores quickly but they do not change service companies (Internet Service Providers, for example) on a whim. Unhappy business customers typically do not decide immediately to walk away from a company particularly if it is a service supplier. This is because services, unlike generic products, are sometimes not as convenient to replace. This is especially true if each individual customer is receiving information-intensive services that offer personalized, variable content or data. Private sector experience both in losing and gaining business indicated that, in the service industry, the loss of customers is incremental both with individuals and within each particular account (group of customers.) The pattern is typically as follows. An individual becomes dissatisfied or unable to access what they need through the services or content their current supplier is providing. The dissatisfied customer begins using an alternate source, perhaps initially on a trial-and-error basis. Eventually, if the competitive service proves to be satisfactory, the loyalty of that customer shifts to the alternate source. If the current supplier's service continues to be unsatisfactory or deteriorates further, then as others within that customer account begin to complain, someone who had already found an alternate supplier makes this known to his/her fellow workers. In sales lingo, this individual becomes a "champion" for an alternate source. After awhile, as more individuals begin using an alternate source, the current supplier's business declines as measured in dollars. Eventually all of the business may be lost.

The good news about this is that it works in reverse, especially if you can convince the customer that you offer a uniquely valuable or highly personalized service that is not easily replicated by an alternate source. If you can satisfy the needs of just one person as a customer, then that person will tell somebody else within their firm and ultimately (perhaps because they change jobs or because of social contacts) will tell somebody else outside the firm. Little by little you begin to multiply the number of satisfactory experiences that particular customer or account is having, and the pendulum swings your way. The competition experiences a falloff in business; yours increases.

The direction of the customer flow, either toward or away from your business, is always determined by the quality of the most recent customer experience. A string of good experiences can be reversed by a couple of bad experiences in a row. A bad experience can be overcome by a series of good ones. There is constant pressure to be consistently good, especially in situations where customer contact is frequent. Otherwise the momentum slowly changes in the wrong direction.

In the service business, one of the most important rules is this; **customers only think you are as good as their most recent experience.**

Shopping for Courses at the Academic Mall

When I was a graduate student, I took a course in how to teach at the college level from the late Norman Cantor, Distinguished Professor of History and Department Chair and later University Provost. Professor Cantor was a great teacher and very popular with undergraduates. He used to warn us that the first lecture or two each semester had better be really good because otherwise you could expect 30% or more of the students to drop out during the first week! Professor Cantor's estimate always struck me as too high, but today many universities semi-officially recognize the first week of classes as a "shopping" period, when courses are tried briefly and assessed for suitability by students. Selecting courses has become a lot like shopping for just the right clothes at the local GAP or Old Navy! The drop rate reported by many courses makes Cantor's estimates seem low. And I thought Professor Cantor was just trying to scare us. It is true that he was a pessimist. His last book was on anthrax and the bubonic plague. Norman F. Cantor, *In The Wake Of The Plague: The Black Death & The World It Made* (New York: The Free Press, 2001.)

Losing Student-Customers

Students generally plan to be at a campus for an entire four years to get their degree. But just because a student-customer decides to take a course does not mean that they cannot be "lost." If the course is unsatisfactory, they have every option not to take additional courses from that professor or even within that department. If a university-supplied service is unsatisfactory, campus housing, for example, students can find alternatives. Moreover, dissatisfied students tell other students why they are avoiding this professor or living off-campus and a ripple effect occurs. The same is true for on-campus services necessary for a good academic experience. If the library is unhelpful or difficult to use, a student may use alternatives (such as Internet search engines) in lieu of more academically oriented research resources and techniques available from the university. Even if that student is unable to have a totally satisfactory experience but nevertheless stays to complete a degree, the ripple effect may still occur when the student leaves campus. First, they may not enthusiastically advise other students to apply to the campus. Secondly, if they are correct in saying that they are not getting what they need and if they go into the job market and cannot find employment that meets their expectations, then that becomes an even more damning criticism of the university. (You might be surprised how many college graduates are critical of their college or university during the interviewing and hiring process, especially if they have had their employment applications rejected a few times.) The net effect of that is that the more capable students apply elsewhere because they are being told that their career aspirations will be better served with a degree from a different university. So an analogous erosion of "business" can occur even though students are often not thought of as customers by professors and administrators.

But once again we are reminded that the process of losing students also works in reverse. By providing unique educational services and a stimulating learning environment and by preparing students well for the 21st century job market, students will tell other students and the university's reputation will grow. The ripple effect works <u>for</u> the university instead of <u>against</u> it. The university can look forward to the latest issue of "The Annual College Review" with anticipation rather than dread.

Having identified the library's customers, we can now turn to the library itself. There are many areas that could be discussed (financial management, project planning, etc.) but our frame of reference is business, which narrows the focus to the two most important areas for a small service organization; developing personnel, which is covered in the remainder of this Chapter and in the Chapter the follows on "Training," and promoting services, which is discussed in Chapter 10.

Developing Digital Personnel

A service business is only as good as its employees and the quality of service they are able to provide its customers. Technology can augment but not substitute for the right people. This holds true for information services as well as for products.

There are two kinds of employee functions in a service business. I have met both types at libraries I have visited. First, there is a face-to-face group (in business these would be called "customer service") that handles communications with customers (students and Faculty.) The staff that performs this function needs to be enthusiastic, friendly and helpful and have a thorough knowledge of the services available. They have to be good communicators and effective trainers. They also "sell the library" by explaining use of the services available, especially new or upgraded services. If they are unhelpful, uninformed and unenthusiastic and if the student or Faculty member leaves with the feeling that it was difficult dealing with the library, then less use will be made of the library in the future. But if the experience is a good one, there will be further use of library resources. The rule for a service business is that you must be "easy to do business with." The same applies to a library.

The best example of a critical face-to-face position at the library would be a reference desk librarian or help desk person, who not only has to have a broad knowledge base but also has to have a wide range of sources at his/her fingertips toward which they can direct students and researchers. Also, they have to have the technical expertise to train students to access this information via either digital or analog methods. Not everyone can do this. But every student or Faculty member helped is a potential "champion" for the library. If the reference librarian does a memorable job, the word spreads; the library system is easy to use and going there gets results. The "customers" will be back.

The second type of employee in a service business is what we refer to a production employee. The same type of person exists in the library. In the digital world these people may be referred to as the "back office." Back office personnel, whether in a business or a library, have to focus on two things, *accuracy* and *productivity*. Nothing is more important than accuracy in a library, whether it is cataloging a new book or physically returning a book to the right place on a shelf. An error can result in a tremendous waste of time and money as well as a terrible experience for students and Faculty. Errors are deadly to a library's relationship with its customers.

On the other hand, productivity (which we can define in simplest terms as the efficiency with which a task is performed) is also important. In a service environment, a key to customer satisfaction is often the speed at which service is available. (It is also often the key to profitability in the private sector.) The mindset for a successful service business is that "if the customers wanted it tomorrow, they would ask for it tomorrow." This is a way of saying that *everything is urgent* when it comes to having products or services available for customer use. There is no tomorrow in a service business, only today.

This puts a lot of pressure on management and employees who have to deal directly with customers who need service *right now*. The sense of urgency that a reference librarian feels when he/she has a student on the phone in need of immediate assistance to complete an important term paper is often not felt in the back office. The need to work methodically and "get it right" is more important to a production worker that any individual student problem. It is the same in a business, where there is frequent conflict between production and customer service over the issues of speed and accuracy. This is especially true in an organization that is striving to provide superior service, where management is raising the bar, setting high standards and demanding that employees meet those standards to improve customer satisfaction. It can be stressful.

Better Accuracy-Greater Productivity

The good news is that a digital work environment can reconcile many of the contradictions between speed and accuracy. In an analog environment speed and accuracy tend to be mutually exclusive, as any craftsman can tell you. But in a digital environment, accuracy and speed go hand-in-hand. Accuracy is improved because data is entered once but stored, retrieved, replicated and transferred electronically (either to customers or from employee to employee) as often as necessary. If the initial data entry is accurate, every replication will also be accurate. Steady improvements in software (automated error checking, for example) and in hardware capabilities (high-speed document scanners, for example) steadily improve the accuracy and speed of the data being fed into the network. Better yet, accessing information using digital equipment and networks occurs at incredible speed. Much more accurate information is available much faster! The stage is set for truly exceptional customer service, far exceeding even the highest standards possible in an analog work environment.

The bad news is that, in my experience, employees can be slow to recognize that the rules change in the new digital workplace. *Simultaneous improvement in both speed and accuracy is counterintuitive to someone who has been trained in an analog environment.* Incredulity or non-cooperation are typical responses. Old work habits, especially in production, fade slowly. It is a long and at time frustrating struggle. Adjustments in managing employees have to be made to overcome hesitation or resistance. Among other things, my business learned to avoid allowing production workers to train other production workers just because they (the trainer) had been there longer. We also learned that job descriptions for production workers needed to be changed continually as digital procedures and productivity goals were refined, and that it was often better to write a rules-based "process description" for a series of workstations than to write a detailed job description for a specific employee or a single workstation. We learned that all process descriptions should contain productivity norms established by management, not by other employees or by any specific amount of work that was traditionally expected of employees. Lastly, we learned patience, because no matter how much training and coaching was provided, old habits resurfaced periodically and benefits were occasionally lost. We "started again" more than once over a period of three years. But when all employees (not just managers and supervisors) were engaged actively in the learning/training process, detection and response to problems was more rapid and damage due to backsliding was minimized.

So why did we continue with the program when it seemed so difficult to get employees to understand it? Because we were excited about the productivity and quality improvements and the effect they had on our business. I have indicated previously that it took 2-3 years for the company to fully convert from analog to digital process. During that time we had to replace half our employees and retrain the other in totally new work procedures. But during that same 2-3 year period, our business (as measured by sales revenue) doubled with no increase in personnel. Once our digital services were stabilized (i.e., significant changes were no longer occurring) sales took another 50% jump in a two-year period, again with no additional personnel. Today, sales are three times higher and the company actually has slightly fewer employees than when we were pre-digital. Most importantly, our customer base is stable.

In my judgment, the reason for such significant growth during that three-year period was the tremendous improvement in the speed and quality of our service. We then encouraged our customer service personnel to raise customer expectations, promote new services as they became available and then to surpass customer expectations by constantly accelerating response times. Our performance benchmark was to routinely met or exceeded customer expectations. Our business grew rapidly as a result. The reason that we did not need additional employees was the substantial increase in back office and customer service productivity. As a result, productivity norms were steadily increased. Since many of the tasks actually became easier, our employees were not threatened by increasing productivity norms. Most importantly, few employees felt threatened by job loss despite wholesale changes in job functions because they could see a steady growth in the amount of work to be done (a function of growing the business.) Nothing soothes the job loss anxieties that inevitably accompany change like having more and more work to do. This is not what management normally expects from labor but, in our experience, this is exactly what happened. It seems to me that this could occur in any organization that aggressively crosses the digital divide.

Was this hard on some of our employees? Yes, but they also enjoyed huge benefits. For example, in the immediate aftermath of September 11, the company committed to full employment and a 40-hour week for all employees, and we fulfilled our commitment completely. High digital productivity made this possible. Our competitors, in many cases mired in analog process, fired scores of people in the mad panic after September 11, ruining both lives and businesses, some of which have never recovered. The same thing could happen in any economic slowdown. Underutilized employees are always vulnerable. To me there is no question that the effort paid off for employees as well as management and I think our employees would agree. And lets not forget our customers, who benefited from our business-as-usual strategy after September 11 and who continue to receive superior service years later.

What tactics and techniques are available to help an organization improve accuracy and productivity? There are many. But they are not as complex as they may seem. Let's take a look at an international productivity leader and find out how they increase efficiency, improve customer service and increase the value of their business, all at the same time.

Wal-Mart as a Productivity Model

Everyone seems to dislike Wal-Mart. Everyone, that is, except Wal-Mart customers and stakeholders. They are thrilled at the low prices, wide variety of products, huge inventories, fast service and (for the stakeholders) excellent profits.

Strategy

On the surface, Wal-Mart's strategy is straightforward; they try to have the lowest price on almost any product they sell, which includes everything from baby food to power tools. Their target market is blue collar and lower middle class shoppers, for whom the big malls and specialty retailers are too fancy, expensive and far away. About 25% of Wal-Mart customers do not have checking accounts or credit cards. They pay in cash.

Wal-Mart statistics, often repeated, are nonetheless mind-boggling. It is the largest retail company in the world, with over 4000 Super Centers. U.S. sales are right at one quarter *trillion* dollars per year, which accounts for 8% of total domestic retail sales and 2.5% of GNP. This is approximately 50% more than Kmart, Sears, Costco and Target *combined*. Wal-Mart is the world's largest private sector employer with 1.3 *million* people on the payroll. It serves one hundred *million* customers each week. Wal-Mart purchases one *billion* dollars worth of real estate each month, most targeted for future Super Centers. In both 2004 and 2005 a new Wal-Mart store will open every day in the U.S plus a few hundred more overseas. It has twice as many stores as Target, its next largest rival. Wal-Mart sales are currently growing at an average annual rate of 15%, which means they double every five years. Wal-Mart says no end is yet in sight, although competitors are everywhere.

When a new Wal-Mart opens, the business dynamics in the area change immediately. A retail construction frenzy results. All the other national chains want to cozy up to the "big box." Smaller local merchants, who often provide a narrower range of goods at much higher prices, are lucky to survive. Even larger stores, especially large grocery stores, undergo gut-wrenching changes. Recent studies show that grocery prices, even in competitive markets served by regional and national chains like Kroger and Publix, decline as much as 35% when Wal-Mart opens a Super Center. Wal-Mart is now the world's largest grocer, with 20% of the U.S. market. Wal-Mart's market share of basic commodities consumed by the average American is enormous; it sells 36% of all dog food, 32% of disposable diapers, 30% of photographic film, 26% of toothpaste, 21% of over-the-counter pain remedies, and 15% of single copy magazine sales, all at lower prices that other retailers are quick to match. Pro-Wal-Mart consumers say that when a Super Center opens in their town prices drop so much that its as if everyone got a pay raise.

Tactics

Low prices are a superficial explanation for Wal-Mart's success. If you shop at Wal-Mart (which I do occasionally) and you pay attention to the price of what you are buying (which many shoppers do not do,) you discover that Wal-Mart prices are not always the lowest. The real explanation for Wal-Mart's success is the "Wal-Mart Way," which combines a *high-energy workforce* with a relentless focus on *productivity improvements* (using digital technology.) It is a pretty picture from any perspective, and it offers other businesses and organizations a model that should be studied in detail.

1. Service-Oriented Employees

First, in regard to employee motivation, Wal-Mart insists that employees focus on customer service above all else. This has two major components; first, *know the products* so questions can be answered quickly and accurately, and second, *engage the customer* on the retail floor so as to assure that they find what they want with minimum difficulty. The focus on helping customers locate products allows Wal-Mart to cram more items onto their retail floors, which boosts sales per square foot. But this strategy assumes the customer can find the items he/she is looking for, which becomes an employee responsibility. Employee attitude is institutionalized in a myriad of rules and procedures. Some, such as singing the Wal-Mart "fight song" just before opening each day, are designed to build *esprit de corps*. Other work rules focus on excellence in service. My favorite is the rule that if any employee comes within 10 feet of a customer he/she is expected to make eye contact with the customer, approach the customer, greet the customer verbally and ask if the customer needs any assistance. Of course this does not always happen, but compare this with the averted eyes and disappearing act perfected by typical retail employees in most stores. Sometimes the Wal-Mart people won't leave me alone so I can shop!

Wal-Mart is not a pioneer in trying to offer consistently exceptional customer service using highly motivated employees. Japanese companies, global leaders in innovation and productivity, have used corporate "fight songs" for decades to pump up their people and they have successfully imported this custom to America. In the U.S., leading high-end retailers such as Nordstrom and Neiman-Marcus are experts at creating a service environment where customers feel comfortable and to which they are eager to return. Wal-Mart has simply extended this level of service to a whole new class of customers. The response has been tremendous.

2. Single-Touch Productivity

Wal-Mart's success is not solely determined by in-your-face service. Behind the scenes, a revolution in productivity has occurred, made possible by a total commitment to digital information. This is called "single-touch" or sometimes "one-touch handling" process control.

Wal-Mart has as a goal that every product they sell will be touched only once by human hands before it is handled by the customer making the purchase. Typically, this one-time touch occurs when the product is taken out of the bulk-shipping carton and placed on the display shelf by a Wal-Mart employee. Every other stage of the process (called supply-chain management) of getting that product onto the shelf, which would vary for each product but could include design, manufacturing and packaging (by the supplier,) ordering, shipping, inventorying, paying, etc. (by Wal-Mart) is to be performed by computers or, where appropriate, automated package handling equipment (managed by computers.) The result is a dramatic reduction in the total cost of the product, including allowance for inventory shrinkage and damage from handling. This is how Wal-Mart is able to sell at such low prices. Their superior productivity in getting the product to the shelf drives their costs (and those of their suppliers, who are expected to cooperate fully with this program) downward and enables Wal-Mart to sell many of their products at prices just slightly below the competition and still make a good profit. This is also why their employees feel so confidant about their company; they know that Wal-Mart can service their customers. It is great for morale as well as profits.

Recent studies of productivity growth in the U.S. credits Wal-Mart (and its suppliers, who are forced to adapt to one-touch requirements,) with about 25% of the nation's total productivity improvement in the 1990's. The run-up in productivity is often cited as a major reason for the healthy, inflation free economy we enjoyed during that period.

Nor is Wal-Mart resting on its laurels. Instead, it is now extending this one-touch strategy to the complete product life cycle by requiring suppliers to imbed a tiny microchip inside of every item. Called RFID (Radio Frequency Identification,) this technology will replace traditional bar code scanning that requires line-of-sight access to every individual item by a person. RFID's emit signals that are read constantly by electronic readers. The data is fed into the company's fully integrated digital inventory system, so the company will have precise information about the whereabouts of the item at any time. Once the customer selects the item for purchase and removes it from the shelf, the RFID system will automatically scan it as it passes through checkout and create a credit card (or cash) receipt, all without any human interference. Even cash will be accepted and change returned to the customer by automated equipment. Pilfered items tucked under a shirt can be tracked out into the parking lot to the specific vehicle where they are hidden. Wal-Mart set a deadline — January of 2005 — for their suppliers to come into compliance with their RFID wireless inventory tracking technology requirements. Company officials expect this change to "supercharge" an already efficient inventory and sales tracking system.

To summarize, Wal-Mart's strategy is to have the lowest price on almost any product they sell. They are able to accomplish this first, by creating a highly motivated, customer-oriented workforce and second, through the relentless pursuit of productivity improvements in their businesses processes. They have achieved tremendous results during the past decade and they have become a role model for businesses and organizations ranging in size from sole proprietorships to major corporations.

9. Training

The management task then, based on the Wal-Mart model, is to intensify and accelerate the information exchange processes of library employees as they work with Faculty and students, their customers. This means building a customer-oriented workforce that provides excellent service due to high digital productivity. How complicated, expensive and painful is the transition to an all-digital information business? Specifically, how are employees trained and motivated to adapt to this changing world and how do we make that adaptation permanent?

The Multiplier Effect

Again, let me refer to my experiences with a service business in transition. To achieve the goal of changing our work force, we used a training and motivation strategy called the "multiplier effect." We can define the multiplier effect (also called multiplier training) as the horizontal transfer of changes in information and work process from person-to-person within an organization. It is well suited to basic changes that require both individual employee training in new technologies and changes in group work habits to accommodate these technologies. To accomplish this, it takes advantage of the global commonalities that have been built into software (and related hardware) over the past two decades (I discuss this further in the section entitled "The Graphical User Interface," below.)

The benefits of multiplier training include,
- Relatively low cost,
- Little or no requirement for additional management or professional resources, and
- A tendency for the process to become self-sustaining and self-reinforcing over time.

Organizations making major changes with limited resources should consider this training strategy carefully but, like Wal-Mart, they must plan carefully and then make a long-term commitment to assure completion.

The multiplier effect is intended to bring about permanent change. Once started, the impact of these changes can grow exponentially. When this happens, process change becomes self-sustaining. Over time, work methods, habits, productivity, etc., are completely transformed. All employees in any given work area are constantly exposed to the process, and the best employees automatically become the leading role models (the trend-setters who validate the changes and the go-to people if there are questions.) There is constant reinforcement of anything new that has been learned and immediate "help" from a fellow-worker if a problem occurs. Assuming each employee has the ability and willingness to learn, all can eventually function at more or less the same level, which we found is generally just slightly below the performance level of the best worker in the group. Most managers would be pleased with these results.

From a traditional educational/academic perspective, the techniques of multiplier effect training are counterintuitive. This is because the training model used was developed for a new kind of software introduced in the 1980's. Because of this, we had to throw out most of what we had learned about training in order to train our work force. This was a different way of thinking about employee training and development, or at least it was for me. After leaving graduate school, I had spent several years of my business career as National Training Director for a Fortune 500 company. In this role I oversaw the development of a top-down technical curriculum complete with a college-like "campus" in Atlanta, video presentations, detailed training texts and a fully credentialed instructional staff. We trained (and continually retrained) our entire national sales staff on our complete product line. I was confidant that I knew how to train.

But when it came to converting a service business to an all-digital format, it didn't work with our employees, despite seminars, instructors and manuals supplied by vendors, paid training supplied by outside "experts," and our own best classroom efforts. We wasted a lot of time before deciding to take a different approach.

Changing Spaces

The advantages of the multiplier effect can be put to best use when converting personnel from a traditional information environment, where employee mindset, work flow and productivity are determined by analog processes, to a digital environment where work flow is computerized and resulting information exchange practices and productivity improvements become the new standard. The goal is to build digital skills and higher productivity into the staff's *normal work habits*. This means breaking old habits and resistance can be strong.

An excellent way to break old habits and set the stage for future change is to make some visual or tangible changes early on in the process. During the conversion of my business, we changed many work areas by reorganizing space and installing new pieces of equipment, software, furniture, etc. We learned immediately that, while "new stuff" is not as important as new ways of doing things, nevertheless new stuff had a powerful impact on both employees and customers, especially during the early stages of the program. Each new installation was a great management opportunity to promote the idea of change and it was staged as a Big Event to drive home this point. This assured that we got the message across to employees that things were changing and they had better pay attention. As bottlenecks and problems with the "new thing" occurred (very slow adaptation to new software, for example, or low utilization caused by lack of training) they had to be dealt with quickly and decisively. But we found that if management is enthusiastic and focused, this begins to create awareness on the part of employees that things really are going to be different. The word spreads, and the next "new thing" is taken a little more seriously; adaptation becomes the path of less resistance. Over time, both employees and customers come to accept change as normal. New hires and new customers think it was always this way. "New stuff" helps the multiplier becomes self-sustaining.

Changing Spaces at Binghamton University

An example of "changing spaces" to signal the onset of much more sweeping changes can be seen at Binghamton University. The university has reorganized the way its students access information by establishing highly-visible computer "pods" on campus. It is an excellent start, significant for students, Faculty and library and computing staff. It signals not only new ways of transferring information but also new ways of teaching and learning. More and better pods will multiply the effect by enabling more Faculty and students to make use of these types of resources. More and better digital content will make pods more and more useful. Eventually, pods will be the norm; carrels and traditional work-study desks will dwindle in number. The university will be dotted with computer pods (and blanketed with wireless networking) for access to and distribution of a large and growing selection of digital resources and information, publication of original works, exchange of resources with the local community, etc. This is a truly dramatic change from the traditional campus study environment. Best of all, it is irreversible. There is no going back.

Peer-to-Peer Training

As employees (and customers) are becoming aware of and comfortable with the idea that changes are coming, the next steps in technology training can be begin. The first stage is for management to analyze work process flow (using flowcharts, for example) and then *revising job descriptions* for job functions that are either not being done at all in an analog environment (example; high-volume document scanning) or are being done but via analog methods. Note that what I mean here is not the "job descriptions" of individual employees but rather the job descriptions of each workstation or group of workstations involved in a specific process, regardless of which employee works at these stations. This is a subtle but important difference. The job descriptions have to be revised in such as way that they take into consideration the total change in workflow that occurs in a digital environment, from the beginning to the end of the process. As each job description changes, others change as well. It is also important that each job description captures the productivity improvement opportunities inherent in digital information processing (it is important that productivity norms be determined by management, not employees.) Once preliminary flow charts and process descriptions have been completed, multiplier training for employees at all levels can begin.

Although help in learning the new processes is usually available from equipment or software suppliers, computer skills training multiplies best when using the *each-one-teach-one* method (peer-to-peer training.) This method is used for many types of training, literacy training for example, when resources are limited or institutional training infrastructure is lacking. The ideal setting for multiplier training is not a classroom or large group, but rather two people at a single cubicle or workstation, each with a computer terminal or PC. This creates a workspace where direct peer-to-peer communication can occur. Initially, only the best, most committed and most digitally literate employees should be used to train new employees coming in or existing employees who have to convert from analog to digital. Seniority is not a criteria used for the selection of trainers. Patience and superior knowledge of process are much more important. But once any employee becomes comfortable with a new process, procedure or software package, that employee becomes a trainer for the next person who needs to learn that skill. As trainers multiply, training occurs continually until all are trained. Also, in many cases job descriptions need to be revised as training evolves because they change during the transition to digital. This is sometimes difficult to do in advance. Until someone learns the new machine, learns the new software, learns how to explain how to use new digital processes or digital information resources to students and Faculty, etc., there really is no way to write a complete job description (remember, we are writing for the process, not for the person.) Practical employee experience brings new insights as to what processes are the best. Management encourages suggestions from employees (and also from customers, through user surveys.) To assure that the right balance is maintained between, for example, speed and quality or simplicity and complexity, management reviews these suggestions continuously. Once a job or process description is finalized, however, it can be blended into the training program and it becomes a new standard for any employee performing that job.

A "Gooey" Interface

The question that should be raised at this point is; how can untrained (inexperienced) trainers working with other employees in an informal environment be effective in training others to use software for processing information? The answer has to do with the user interface that has been developed over the past 20 years and has become accepted virtually everywhere on the planet. It is called "Gooey" and was pioneered by Apple Computers.

Training the Apple Way

1984: Orwell Was Wrong

Nineteen eighty-four was the most important year in the history of personal computers. That also happens to be the year I purchased an Apple Macintosh system. The "Mac" was one of the first ever to be sold with a Graphical User Interface (GUI, pronounced "Gooey.") Nothing in the world of computers was the same after that. Orwell could not have imagined how the year he made so famous would ultimately be remembered.

The first Mac consisted of an 8mhz CPU with 128k of Random Access Memory (RAM,) a single floppy disk drive and a 9-inch black-and-white screen. There was no hard drive. Initially you had to remove and insert floppies using the single disk drive in order to load programs or save data. It was incredibly slow by today's standards. In-and-out, in-and-out, one could swap floppies many, many times just to save large files or load a program. Later Apple introduced an external floppy disk drive, but there was still no hard drive and the floppies were limited to 400k. I remember having so many floppies that I purchased a plastic disk organizer for my desk just to control the mess. There were hundreds of gadgets like this developed just to support the Mac. In its early days it needed all the support it could get.

The Mac's most unusual hardware/software feature was the mouse, which enabled the user to connect to an on-screen point-and-click interface. This allowed commands to be entered into the computer without the need for the much slower and more cumbersome command structure of MS-DOS (Microsoft's Disk Operating System) and other similar operating "languages." The mouse interface had been introduced previously as part of an Apple system called "Lisa." Lisa was a total failure insofar as market acceptance was concerned and the mouse was cited as one of the reasons. It was dismissed by many computer experts as a gimmick or a toy, not a device that people would use if they were serious about computing.

The original Mac package also came with a dot matrix printer. It was slow but reliable. Shortly after the original Mac was launched, Apple introduced the LaserWriter printer. It was remarkable because it contained a large number of fonts in many sizes on a Read Only Memory (ROM) chip. This was a tremendous breakthrough in document formatting and design. Initially only Mac Write word processing software (which came with the system) was available but soon Aldus introduced "PageMaker", and the Mac spawned an entirely new industry called desktop publishing (DTP.) It seems like ancient history now, but for a decade or so DTP businesses opened up near every office park in America, replacing the slow and expensive "typesetting" services offered by offset printers. A newsletter, report or brochure that previously took weeks to typeset and print could now be done in a day or two using a Mac and a high-speed copy machine. Tens of thousands of traditional print shops went out of business. Tens of thousands of new quick copy shops riding the DTP boom were opened. The Mac became an engine of economic growth and transformation. And, as the price of the Mac and the LaserWriter declined, businesses could afford to do desktop publishing in-house as well. The quantity and quality of printed communications increased exponentially.

Learning How to Use a Mac

One could tell immediately that something was different about the Mac by the way it was marketed. First of all, the only advertisement for the Mac, which appeared during the January 1984 Super Bowl, had a surreal science fiction motif that combined images of regimentation and rebellion. It later became famous, hailed as a dramatic breakthrough in advertising communications, subliminally suggesting the empowerment of creative individuals against the powerful and the entrenched. At least that is what the ad experts said it meant. It had little information about the actual product and no information at all about price. Later in the year, a 16-page brochure was released, but only after Microsoft had introduced their own GUI Operating System called "Windows" (which Bill Gates originally wanted to call "Interface Manager.")

The Mac ad must have been effective, because I decided to go take a look for myself (the ultimate tribute to an advertisement!) At the time, I owned a small manufacturing business and any new ideas or gadgets I could find might help.

I thought the Mac would be sold at Radio Shack or one of those small computer stores that had opened to sell the Atari's and the various computer kits that had come on the market. But the venue selected was Rich's Department Store at Lenox Mall, the largest department store at the largest mall in the Atlanta. Rich's was known for clothing and house wares but the most sophisticated electronics they had ever offered were TV sets. Sure enough, the Mac was sold in the TV department.

I went on a Saturday so I would have plenty of time to test the system. There was a line waiting to get in. They had a Mac set up on a bare table with two chairs, one for the customer, one for the instructor who would give the demo. There was no hype, not even any posters or brochures available. It couldn't have been more Spartan. Another sales rep explained the process; each customer would have a total of five minutes for the demonstration, no more because of those waiting in line (so much for spending my Saturday with the Mac.) The sales rep would hold the mouse and do a brief demo that would take about three minutes after which the customer could hold the mouse and "play" on the screen for an additional two minutes. At the end of five minutes the demo was over and you had to leave to make room for the next customer.

You could learn almost nothing about the system in five minutes. Everything was totally new and disorienting. The sales rep did a largely silent demonstration by using Mac Write to type a sentence onto a blank "page," name and save the "document" and then delete it by "grabbing and dragging" it to the "trash" on the "desktop." He then turned the mouse over to me. I didn't know what the "desktop" was, didn't know how to "select" an item, and was expected to master the Olympian athletic skills necessary to "grab, drag and drop" and suddenly the demo was over. Next person, please.

I ordered one on the spot. But as soon as the paperwork was completed, I was informed that there was no inventory! Rich's had not ordered in a single unit. Everything would be shipped from Apple in California on a per order basis. Lead-time was about six weeks. Later it came out in the newspapers that there was no sales forecast and no one had a clue as to how many (or how few) units might be sold, so they didn't order any at all. If you wanted one, you would have to wait. (Apple sold only 50,000 units nationally during the first two months, a bitter disappointment to their marketing staff. They considered the product introduction a failure.)

The price was $3,600.00 and change. This was a large amount for me at the time and I decided to charge it on my Rich's credit card. More surprises. Rich's wanted 50% in cash or charged to a credit card other than Rich's before the order could be placed. The balance would be charged to the Rich's card when the product was actually available. Later Rich's explained that, because of the totally unknown user acceptance of the technology, they feared many orders would be cancelled or returned or kept but not used and therefore not paid for so they decided to force consumers to pay part of the bill in advance and also spread the financial risk to Visa and Master Card. (TIVO had the same problem a couple of decades later. A lot of TIVO sets were sold when it was first introduced but a large percentage of those were returned because consumers could not figure out how to integrate the hardware with their existing TV set and could not learn to use the on-screen programming. Partly because of this TIVO has struggled financially, despite having a hugely popular product.)

Just about everybody underestimated the impact that Apple's GUI technology would have on the world. I wrote them a check and waited. As it turned out, it was a good decision.

The Graphical User Interface (GUI)

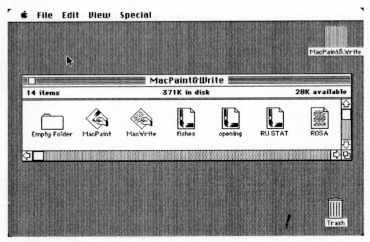

Typical Screen Shot From the Original Mac Operating System 1.1

The Mac came with two software programs, Mac Write and Mac Paint. Each created a sensation, not just because of specific features (Mac Write, for example, was the first word processing program to allow instantaneous global formatting of a document) but also because of the short learning curve and ease of use. The computer itself came with a thick manual, each software program also had a thick manual and even the dot matrix printer had its own thick manual. Each was individually shrink-wrapped. I hadn't seen any manuals at the sales demonstration and it was intimidating, almost depressing, when the cartons were opened and I discovered all of this extensive documentation.

The first manual I opened dealt with basic system operations. There was page after illustrated page of how to drag, how to drop, how to open files, how to save files, how to access drop-down menus, how to delete files, how to reposition folders on the desktop, etc., etc., etc. There were whole sections that did nothing but define icons, some of which changed shapes or "faces" as you used them. There was happy Mac, sad Mac, puzzled Mac, the desktop "Trash" basket and scores of others, each a unique graphic of one type or another. Hence the phrase Graphical User Interface (GUI.) But after awhile I learned from the manual that, by moving and clicking on these icons, I never typed anything that even remotely resembled a command except perhaps when I named a file. Everything was done by point-and-click using the mouse and the icons on the screen. Better yet, *the same icons were used in every software program*, so that once you learned the meaning of an icon in Mac Write, you could be fairly sure that the same icon would have the same meaning in Mac Paint. After a little experimentation, navigating through a new program was intuitive; you just sort of "knew" how to do it because you "knew" what the icons meant. This was a fantastic breakthrough in computer training.

The rest of the manuals never got opened. There was no need. Over the years I have spoken with many Mac people who still have those manuals in their original shrink-wrap. Learning and using the software was so simple that is was a waste of time to thumb through hundreds of pages of documentation. It was easier just to click and try something and if it didn't work, click the "Undo" command and try something else. As time passed and learning the Mac became easier and easier, I realized what the demo had been...an ideal way to train a newcomer on the Mac. The instructor said little during the demo and then silently faded into the background as I tried my hand at the new interface. I only had two minutes and accomplished very little, but if I had been given two hours with an instructor at my side answering questions and inserting prompts as needed, I could probably have learned Write and Paint with time to spare. As it was, I learned them in a couple of days completely on my own, with no training at all. In all the years I owned Macs, I never once had to call a Help Desk for assistance. I would just ask somebody else with a Mac and, presto, they would almost always be able to help me figure it out. I returned the favor whenever I could, and a peer-to-peer "support group" of my Mac friends was born. We did not need outside training or help; we could do it ourselves.

Today, few software programs have manuals and those that do offer them online as a no-cost .PDF download option. Almost all software is now learned the way I learned the Mac in 1984. I recently visited one of those new retail stores that Apple has opened "at a mall near you." I noticed that they had about twenty stand-up computer workstations, each with a modern Mac (G4's and G5's.) Customers can "play" with a Mac pretty much as long as the like, mousing through all the basic operating system features and applications. Mac employees lurked in the background to answer questions, but did not intervene unless invited. Customers were teaching themselves the new Macs (and making buying decisions in the process.) I asked one of the employee-lurkers about manuals; he said matter-of-factly that they could be downloaded but that there was really no need. "The software is completely intuitive."

We tend to forget today how striking the contrast was between GUI and DOS. DOS was like a foreign language. It has precise grammar and syntax much of which had to be memorized. Like a language, there were volumes of information (the equivalent of dictionaries, thesauruses and Turabian-like manuals) in book form. Most people using DOS built up a library of printed resources to help them along. Also like a foreign language, if you didn't use DOS regularly, you soon forgot how. It was not like riding a bicycle, easy to pick up again after being away from it for awhile. DOS could be learned in a classroom, almost like math. And DOS required typing skills that were either very good or very slow (so as to avoid typos.) Typed errors, no matter how minor, could be fatal to DOS. Prior to GUI, there was concern that DOS required a special type of intelligence and mathematical skills on the part of those wishing to use computers and that the rest of us might become a secondary class of technology-impaired citizens, dependant upon a special priesthood of computer programmers and operators. With GUI, computer use was democratized. If you could see and if you could move your hand in a semi-coordinated way, you could use a PC, even if you couldn't read (today there are GUIs even for the blind and disabled.)

A Future Without Words?

Literacy (meaning here the ability to read a written language) is considered one of the most elemental skills in the world. Functional illiterates are often incapable of earning even a bare subsistence living in modern society. National and even global campaigns are launched periodically to eradicate the scourge of illiteracy. But, it is interesting to note that, in the past two decades, a growing amount of information usually communicated with words and text is now being communicated graphically. This is a global phenomenon and has created momentum toward a universal graphical language understood by all without the need for training or formal education. To take a simple example, where there used to be a traffic sign that said "No U Turn" there is now a universally understood graphic symbol. No further explanation is required. Everybody who navigates on an open thoroughfare (car, boat, bicycle, etc.) knows what it means. Moreover, once somebody understands the No U Turn icon, they need no training at all to understand other traffic related signs. Drivers understand intuitively even if they have never seen them before, which is remarkable when you consider that misinterpretation could result in a fatal accident!

Anybody who understands any one of these icons intuitively understands the others as well.

Or look at these International fabric care symbols and think about the process of washing, drying and ironing a shirt. You will probably figure out what they mean. But even if you have to ask somebody what one dot means (low temperature) you will probably not have to ask what two or three dots mean. You will know intuitively. This is the advantage of graphical communications. There is no learning curve.

In the past few years GUI's have evolved the point where they have become a mainstream digital communications tool, not limited to any specific software or Operating Systems or even to any specific digital technology. A case in point is *emoticons*, which are used to convey subtle thoughts and attitudes. Emoticons can be defined as facial expressions made by a certain series of keystrokes. They are most often produced with the image of a face sideways. They started out as a form of digital humor but have become an integral part of online communications, especially chat rooms and email. Yahoo, for example, offers a complete online emoticon library as part of its Instant Messaging Service. Emoticons have now spread to cell phone and PDA (Personal Digital Assistant) text messaging. In a real sense we are witnessing the developing a visual language (there are no specific sounds associated with any of the emoticon faces.)

Emoticons are the direct descendants of the Sad Mac icons popularized by MacIntosh in the 1980's. Some simple examples are: :-) [happy] :-([sad] :-< [very sad] and so on. At the basic level, you would probably figure out most emoticons intuitively after having one or two explained by someone already familiar with them. But emoticons are not limited just to simplistic feelings or concepts. They are becoming a subtle and sophisticated tool for expressing complex ideas and opinions. There are online dictionaries that track emoticon usage and the changes in definitions. We are watching a whole new form of high-speed language evolve which requires little or no training on the part of users. Compare that to the fictional Klingon "language" that has been developed by Star Trek enthusiasts. According to the online Klingon dictionary, the English number sequence **1234** is translated as **wa'SaD cha'vatlh wejmaH loS**. This non-graphical language is moving in the wrong direction, that is, toward extreme complexity and non-intuitive vocabulary and grammar, exactly the opposite of GUI-inspired language. It is DOS vs. GUI's all over again. Humans will likely never speak Klingon but we will probably consume ever-increasing amounts of information through wordless graphics.

Some of the more complex ideas conveyed by emoticons include]? [moving away and wondering about you?] >:-(8 *>-[Has a disparaging opinion of feminists.] To the graphically illiterate, these are unintelligible, but to the text-messaging generation, they make complete sense.

Symbols may also be the future of machine "literacy." Human ability to understand symbols or at least intuitively interpret new symbols will be essential if machines and people are to work together in a synchronized manner. For example, a team of Australian scientists and engineers, attending a recent International Conference on Intelligent Robotic Systems in Japan, demonstrated a car that can read road signs and alert drivers who ignore them or fail to respond properly. The team found shape recognition to be the most reliable system for picking out signs against a background of objects visible from the automobile windshield in varying light and weather conditions. Previously, attempts at automatic sign recognition were based on the detection of colored patterns, but symbol recognition was more reliable. This will add to the tendency of roadside traffic information to be displayed with internationally recognized symbols understood both by people and machines.

To summarize, since the mid-80's communication among computer users and information workers has been greatly simplified by the advent of the Graphical User Interface (GUI.) The use of GUI's continues to grow and is no longer limited just to computer software and the Internet. It is spreading to casual interpersonal communications and other information exchanges. Although still evolving, the use of GUI has transformed the way in which workers learn software, greatly simplifying and accelerating the process for each new generation of software products. Information intensive organizations need to make the necessary adjustments in their worker literacy as their services become increasingly digital.

Accelerating Change Through Turnover

A certain amount of turnover can be a good thing during the transformation of an organization. While interviewing and hiring is time consuming and often nerve-wracking, it nonetheless creates an opportunity to quickly upgrade employee skills. These new hires can either be tossed into the pool with everybody else, i.e., into a traditional work environment with analog productivity expectations, or placed into work areas where digital skills are critical and digital productivity is expected. The better choice is to hire people who already have acceptable digital skills, put them in areas that are in the process of being digitally upgraded, and then use their expertise to modify work processes and train others in their group. Simply replacing analog people with analog people is a missed opportunity to facilitate change. If positions that need to "go digital" are held by employees whose work experience and skills are primarily analog, the employees need to be moved out of these positions into what becomes a slowly-dwindling supply of jobs where analog skills alone are sufficient. Over time, the balance between digital and analog positions is going to shift. When I converted my business, we lost about 50% of our employees during the change, and by "lost" I mean that, despite reasonable efforts on our part (extra company-paid training, for example,) people who had been satisfactory analog employees had to be replaced because they could not or would not become digitally competent. Most of them eventually recognized that they were not going to do well in the new environment with new equipment, processes and mindset, and that the future with us held limited analog opportunities. They decided to leave the company without even feeling particularly angry about it. What once was a satisfactory work experience had become unsatisfactory and they simply felt better about their employment situation when they went somewhere else. Some, of course, had to be terminated. Overall, it was painful, but not as painful as I thought it would be. Fortunately, we were in a local economy that was producing a steady supply of other jobs, so job change was a realistic option for them. The bottom line is that half of our employees were unable to navigate successfully through a 2-3 year transition and were lost and had to be replaced by new hires coming in from the outside, applying for and being hired for jobs with descriptions that made it crystal clear that if they were not digitally capable there was no chance that they would be hired.

Employee Turnover in Academic Institutions

According to recent government statistics, 7 million workers in the US leave their jobs every 90 days and 7 million workers are hired into new jobs. That is approximately 5% of the total US workforce. This is by far the highest job turnover rate in the world. According to a recent report cited by a major Eastern University, 17% of employees at US colleges and universities turn over annually. But when it comes to professional librarians there is currently a shortage. The median age of librarians is 47 years. This is the highest of any of scores of occupations surveyed. This is in part due to the fact that many librarians are pursuing second careers. Nearly 58 % of professional librarians will reach age 65 between 2005 and 2019, according to the 1990 census. More than 140,000 librarians work in public, school and academic libraries. Median income is $34,000.00. In 2003 the federal budget allocated $10 *million* to train and recruit new librarians.

Service Company Hiring Profile

People are the key to good service regardless of the level of technology deployed. In the 1980's Eric Toffler and others suggested that, as the level and sophistication of technology increases, there would also need to be an increase in the quantity and quality of human interface with the "technology-enabled process." This is the best way to assure effective communications and interaction both among people and between people and machines. This was the now famous "high tech/high touch" construct which, when first proposed, was not well understood. But two decades of technological breakthroughs have proven the point. We now understand it intuitively. In any business, but especially in an information intensive business, who you hire really matters because what they *do* really matters to customers, regardless of how automated the work environment or digital the content may be.

But how do you know who will become a good employee? This question vexes every business owner and manager. I used to ask my counterparts (other business owners or managers of comparably-sized organizations) how successful they were at hiring, that is, what percentage of their hires prove to be suitable and also remained on the job long enough to make the effort and expense of hiring them worthwhile. The high end of the range was about 50%, that is, about half of those hired. This is discouraging. And I cannot say that, in my early (analog) years, I did any better. I fell into the trap that most employers do, hiring based on directly relevant work experience. Either that or I hired people that looked and acted like other employees who were doing a good job, as if that would somehow help. (Hiring look-alikes is far more common than you might think in business, even though everybody agrees it is the epitome of bad management. When I worked at a Fortune 500 Company and was responsible for training newly-hired sales and technical reps, we made a game of guessing who a new-hire worked for by comparing how he/she looked to pictures of managers in the company. In some instances, our accuracy rate approached 80%. We tied to discourage this practice because, predictably, managers were more reluctant to fire weak employees who looked like themselves.)

But things changed. Once my own company started making the transition to an all-digital information and work environment, I was able to improve my hiring practices significantly. It got easier to screen and hire and I was able to develop a hiring profile that dramatically improved my success rate in hiring. The timing was right, because we had to replace about 50% of our workforce over a three-year period as we made the transition from analog to digital information services. Moreover we have enjoyed a far more stable workforce since the conversion. Less difficulty and expense in hiring and less after-hire turnover make this an idea worth exploring by any service organization whether private or public sector.

We learned to look for two qualities in the people we interviewed. I adopted a slogan to keep me focused:

$$\text{Altitude} = \text{Attitude} + \text{Aptitude}$$

You can probably figure out intuitively that "altitude" is the overall level of performance achieved by the employee after he/she is hired. But what do we mean by "attitude" and "aptitude?"

Attitude

First, let's consider attitude. We learned that a company providing service needs employees that have a "service-oriented attitude." What is that and how do you know it when you see it? It is a personality trait, a mind-set that is expressed in many ways both during an interview and during the first few days and weeks on the job. Some people do not object to being placed in a position where they are expected to provide service to other people. They don't find it personally demeaning and they don't mind going out of their way to assist customers in need. An extension of this is that they feel that it is important to have the knowledge and skills to help customers quickly and accurately. They want to be good at what they do because they think it is important to be of help to customers.

The opposite kind of job candidate expects to be the recipient of services, to be the served, not the service provider. They somehow feel inconvenienced or personally diminished when they cater to the needs of customers. This can quickly deteriorate to the point where the employee views the customer as an obstacle to their daily work, not an opportunity to do good work. They say things like "I couldn't get anything done this afternoon. I kept being interrupted by customers" Or, "This would be a great place to work if it wasn't for the @%$*&! ^ customers!" (I have actually seen this on a sign in an office on more than one occasion.)

How to separate the two during the interview process? Ask open-ended questions and listen closely to the answers. A little experience will teach you which questions work best for you. I have learned to ask some "trick" questions as well as some straightforward ones. For example, I show them one of the "nebbish" cartoons (see the section on "How to Lose Business by Laughing at Your Customers When They Are In A Hurry" in Chapter 2) and ask them if they think humor is appropriate in the workplace to lighten the mood. The question is not about humor; it is about the content of that particular cartoon. I am hoping that the job applicant notices and comments on the inappropriate attitude shown toward customers, so that whether or not they think humor belongs in the workplace, they should certainly be critical of that particular attitude. Or, I create a hypothetical situation (one that really happens occasionally.) At five minutes before closing time our best customer rushes in and says she needs some help right away with a project. She apologizes for coming in so late in the day but insists that somebody stay late to assist her. Most job applicants are able to figure out the answer to that question (somebody has to stay and help the customer) but many change their response when I make the hypothetical customer somebody new that we have never seen before. It takes a higher level of service commitment to treat a stranger the same way one would treat our best customer BUT experience shows that superior service for new customers often leads to future business. If the success of your business depends upon your reputation for giving good service, somebody needs to stay and take care of the customer even if you are not sure who he/she is.

Much of the same information can be gathered by asking the job applicant how his/her former (or present) employer treats customers. Many will complain about all the things they have to do to keep the customers happy. Others will complain about how their former company never did enough of the right things to keep customers coming back. Which one would you rather have on your staff? The answer is obvious.

Aptitude

The second requirement is aptitude, which in the case of our digital information services business means the ability to learn to use basic software packages and local network(s) to gather information and communicate with and/or create content for customers. In the analog days, there was no equivalent skill set to look for (except perhaps good handwriting or manual dexterity) but in the digital world good "mouse skills" should be a minimum requirement. What do we mean by mouse skills? This can be defined as a good working knowledge of one of the more recent versions of Microsoft's Windows Operating System and familiarity with one or more of the basic applications that run on Windows. Examples would be Word or Excel, but there are a lot of others that would serve the purpose. Even if the employee will never use these programs, the fact that they understand how to use them with standard Windows Graphical User Interface (GUI) is a good indication that they can learn new software quickly, thereby adapting to the specific job requirements with minimum training and unproductive time.

We learned to "test" job applicants either by asking them questions about Windows or by handing them the mouse and then sitting back to see what happens (along lines of the old Apple McIntosh product demonstrations I discussed in Chapter 9.) You can tell a lot by watching someone just for a couple of minutes. I would ask about some typical tasks relevant to our business, such as how they would format and save a Word document in preparation for uploading it to a web page. (Answer; the document has to be converted to HTML or some other browser-readable format.) The easiest way is to scan it to .PDF or convert it to .PDF using Adobe Acrobat. No, I say, we don't have a scanner or Acrobat, only Word. That being the case, they should know (or be able to find) that by extending the File menu below the Save As option they will find a Save As Web Page option that creates an HTML format for the document. They also should know (from a few bad experiences) that it is best to create a separate folder somewhere on the local system where the Word-to-HTML can be stored because otherwise the files and folders created in support of the HTML format will be scattered all over the place and impossible to find when it comes time to upload to the web site. If the applicant has never done any of this before but is otherwise has good mouse skills, I start walking them through the process verbally. All they have to do us follow my instructions, which is easy if they have experience with Windows. I can tell almost immediately if they can do it. But if the dropdown menus are confusing, or if they have dexterity issues with the right and left mouse buttons, the interview is pretty much over.

If this all strikes you as radically different from standard job applicant review techniques, you are right. Normally you would read an applicant's references, look for the appropriate work experience, etc. But if the candidate has clashed with her former supervisor over low service standards, the references may not be good or there may be none at all. Or if the work experience allowed an applicant to develop some bad habits in how she treats customers, experience is a liability, not an asset. In fact, in a rapidly changing work environment (converting from analog to digital, for example) experience may be a substantial liability because of the tendency to resist adapting to new works standards or procedures. On the other hand, if an applicant has a good sense of service responsibility to customers and also understands the basics of Windows and a handful of applications running on Windows, you have someone worth considering.

There are millions of workers today who have excellent, if undisciplined, mouse skills. They have learned from grade school on up through college to use Windows and Windows-compatible applications. They have likely spent countless hours surfing the Internet, which itself is great mouse training. This has become as basic as knowing how to read and write. The next time you run an ad in the "Help Wanted" section, try adding the phrase "Good mouse skills required." You will be surprised how many qualified prospects there are out their looking to put their skills to use in an information-intensive environment.

There have been some unexpected discoveries when it comes to hiring customer-oriented people with good mouse skills. For example, we found no difference in the size of the applicant pool or in our ability to attract and hold good employees over a surprisingly wide salary range. There was almost no correlation between performance and salary in the $10.00 per hour to $20.00 per hour ($20,000 to $40,000 per year) range. (This is not so much the case at the supervisory or management level.) We also found that lower paid employees were more likely to stay with us and not move on to another job. We had expected the opposite to be true as lower paid employees sought to move up the pay scale by jumping to another company. It was the higher salaries that moved on the greener pastures. Needless to say we quickly learned to hire at the lower level and bring employees along salary-wise as their on-the-job performance warranted.

Yet another unexpected discovery about hiring service oriented people with good mouse skills. In the $20,000 to $40,000 wage range, those with college degrees performed no better (and interviewed no better) than those with little or no college education. This included applicants with two and four year degrees from "technical schools" like DeVry Institute (a nationwide change of for-profit campuses that focus on technical and business education.) I think the reason for this is that mouse skills and digital communications experiences evolve continuously. Because of this, they are not as dependent upon formal education. They are acquired simply by navigating through a typical 21st century day. High school and even grade school students surf, chat, cell, text message, blog, email and download and burn CD's every day. They carry iPods in their backpacks, camera phones on their hip and have PC's at home. Most are digital communications experts by the time they leave high school. They can step into many communications jobs that require these skills right out of high school and then pursue a degree or other appropriate credentials through local community colleges, private technical schools or through distance learning. The comparative skills of employees with college degrees vs. those with only a high school diploma will be explored in an upcoming season of the popular business reality show, *The Apprentice*. One of the competing "teams" will have only high school graduates. The other team will have all college graduates. They will perform management tasks for four major companies, including Burger King, Home Depot, Domino's and Nescafe. "I have decided to take the series into a new realm," Donald Trump, host of the show said in a statement. "We wanted to see what would happen if we pitted college grads against high school grads." Based on my interviewing and hiring experience over the past few years, if the jobs require a significant amount of digital expertise, some college graduates are going to have their egos bruised by the time they hear "You're fired."

Finally, what about social skills? One of the reasons for a job interview is to assess the personal poise and verbal communications skills of a job applicant. This may or may not be reflected in the resume and the references. These skills are important for people who will be communicating verbally with customers, especially if it is face-to-face. Again, it seems to me that things have changed significantly over the past decade. High school and college-age people and young professionals are much more social and socially aware than they used to be, mostly because the digital age has helped accelerate the process of social maturation and greatly intensified the social environment in which they live. Social communications are the largest daily time allotment made by 21st century teenagers and students, due in large part to the digital technology that makes it possible to have contact with tens or even scores of people each day. Pre-digital, only the socially elite could hope to have this degree of contact and social interaction, and then mostly within their own socio-economic strata and local geographic area. Having a pen pal in another town was considered a significant breakout experience for the average person. Today's information workers are well aware of the benefits of blending their social skills with their digital communications skills to create a self-confidant persona. Add a few trendy togs from Banana Republic or Kohl's (perhaps as seen on the latest music videos on MTV) and you have the ideal service worker, digitally literate, personable and looking good. We found that this is exactly the kind of person customers like to deal with.

Hiring Profile Summary

To summarize, in the past two decades digital technology using Graphical User Interface (GUI's) has spread throughout the world exposing hundreds of millions of people to the basics of business and communications software. This has simplified the task of hiring the right employees to work in an information services environment. The widespread use of digital technology for personal communications and entertainment has also improved the interpersonal communications skills of the average job applicants, even of those with little or no actual work experience or formal education. Older screening techniques, such as resumes, personal references and prior work experience, have become less predictive of satisfactory on-the-job performance. The most elusive quality remains a service-oriented attitude. This may require careful observation (combined with motivational training) during a trial employment period before a final decision can be made about an individual job candidate.

10. Promoting Service

Service is easier to promote than a tangible product. This may seem counterintuitive, especially considering that a substantial part of what constitutes service is the customer's *perception* of the value and quality of the service they will receive. "Good" service is often intangible, occurring only in the eye of the beholder. With a tangible product, customers are better able to see what they are getting and make comparisons accordingly. A service may be helpful, save time and money, enhance business or financial performance over time, etc., but these things are more difficult to measure. This is especially true of information services, since oftentimes the customer can calculate the actual value only after the service has been used (examples being investment services or academic research services.)

It is also less costly to promote service. In fact, as we will see shortly the marginal cost of certain kinds of service promotion, whether face-to-face or digital, can be close to zero (marginal cost being the effort and expense of getting the service message out to one more prospective customer.) BUT, it may take more time to get results from service promotions. Whereas tangible products can be promoted via short, intense print or TV ad campaigns highlighting new product introductions or limited-time price reductions (on sale this weekend only!), service is best promoted by constant, steady visibility campaigns that emphasize regular changes in content or features over time (sometimes called "drip marketing."). Constant promotion is necessary because customers usually respond more slowly to a new service and are slower to switch from one service provider to another (unless the quality of the service they are using deteriorates rapidly.) Compare this to the way consumers change the gasoline they purchase for their car. They may use a different brand every time they fill up, making their selection on price and convenience (where they happen to be when they notice they are low on gas.) But, no matter how slow the response, service must be promoted to prospective customers. As many service businesses have discovered, just because you build it does not mean that customers will come. You have to draw them in.

Buzz and Viral Marketing

There are two strategies that are especially helpful in promoting information services. These are; buzz (also referred to as viral) marketing and push marketing. They are well suited because they capitalize both on the growing customer demand for variable information content and the relatively slow process of customer acceptance of new or different sources for information.

Buzz marketing, in its digital form (using email) is also called viral marketing. Yet another term for buzz or viral is network marketing, which describes a key component of the strategy of both buzz and viral. Buzz marketing is sometimes also confused with word-of-mouth marketing although, as we will see, buzz goes well beyond word-of-mouth. There is a *Viral and Buzz Marketing Association* (VBMA) that attempts to put all of these terms into their proper categories with appropriate definitions. Various marketing consultants favor one term or another, but they ALL have a single underlying theme — *contagion*. The goal of buzz is to make information about a product or service infectious so it will spread directly from one person to the next.

How is this accomplished? In its most effective and exciting forms, buzz/viral marketing takes full advantage of people networks. These can be either a local community of people with shared interests or a more geographically diverse (even global) group who do not actually know each other but who share their common interests over the Internet (via email or blogs or in chat rooms.) At the risk of burdening you with yet another confusing term, I like to refer to this as peer-to-peer (P2P) marketing, which does not mean that P2P Internet technology is required to make it work. The P2P analogy is that, like the computing architecture of the same name, **information flows directly between people and does not have to be processed through or from some central point in a communications network.** This is what we mean by contagion. Viruses make people ill, so perhaps the analogy to a runny nose is not the most appealing way to think about marketing and promotion. BUT you catch it directly from another person, not from some centralized virus dispensary. The value of the analogy is that it helps marketing strategists create messages that spread like a cold virus, from person to person, mutating (changing slightly to become more interesting) each time it replicates, getting stronger (more detailed) and adapting to each new host (person) so as to better penetrate defenses (indifference, for example) break down natural resistance (not wanting to spend money, for example) and becoming more effective in the process. A really successful viral (buzz) marketing message, like the flu, spreads almost without barriers until all potential hosts have been infected.

What are the main components of buzz/viral/network/word-of-mouth marketing/P2P? Several are usually mentioned when defining a buzz campaign. Each can be reduced to a short, simple concept; make it *free*, make it *memorable*, make it *easy*, make it *expandable* and be sure it is *networked*, Not all are appropriate for every kind of product or service in every geographical market area and not all are necessary for buzz to succeed. But some combination is usually present in a successful buzz campaign, and the more of the basic components a buzz campaign uses, the better the chance of success. They can be used by any organization that has developed an interesting product or a useful and professionally delivered service and, because of a limited promotional budget, has more time than money to invest in marketing. Let's look at a couple more in detail.

Make it Free

First of all, it helps if the marketing campaign can offer something for free. The consumer's eye is always attracted to free. But free can be difficult if sampling is not an option. Imaginative strategies sometimes have to be developed to give something away. Let's assume, for example, that you manufacture or sell automobiles. You can't give cars away and it is not always easy to get prospects to visit your showroom. BMW solved this problem by making short mystery movies (with their cars prominently featured along with top-name stars) that can be downloaded from their web site. This is not great cinematography, but it provides a few minutes of pleasant diversion (most down-loaders are at work and are supposed to be doing something more productive!) Most importantly, people can email the web page link to all of their friends and co-workers, spreading the message quickly. Since BMWs are sold and used worldwide, the Internet is the right medium because the product has no geographical limitations. Information (entertainment) is used to draw attention to a tangible product (automobiles) but, unlike the product, the information can be given away free to all who are interested.

Purina dog food had a different problem. First of all, not everybody likes dogs so they are not going to be interested in dog food. Secondly, shipping free samples of yummy dog food all over the world is impossible. But a free, animated "virtual dog" that can be downloaded and placed on a PC desktop or imbedded in the "skin" of an Instant Messaging Service (IMS) screen can put the Purina name in front of dog lovers everywhere. All they have to do is download it and, in fact, the Purina virtual pooch has been a big hit. The mutt can be "played" with and even taught some virtual tricks and, like the BMW link, enthusiastic pet owners can email the web address of the "virtual dog pound" to friends. Cat lovers can ignore it. In a similar fashion, Burger King's viral chicken sandwich promotion (the

"Subservient Chicken," someone in a chicken who responds to typed commands) had about 14 *million* visitors to the web site in the first two months of 2005.

Services that provide personalized and/or variable information generally offer more options for sampling. Information that constantly changes can be given away at little or no long-term cost and can become a very effective component in a buzz marketing campaign. Most variable information has a shelf life of some duration and grows stale over time. Therefore, a free sample of variable information has real but temporary value. This allows the prospective customer to experience some actual short-term benefits from having the information available. The temporary value of free information is a powerful multiplier for a buzz message and, as we will see below, can also be combined with push marketing to create an effective promotional program. Free information is also risk-free from the user's point of view, which in a business environment may be just as important as cost free. There is no requirement to abandon current information sources nor is there any financial commitment. Nobody else even has to know that you are sampling a new source or experimenting with it, not even your boss.

Let's look at a hypothetical example of how buzz marketing works. The buzz strategy is to induce individuals to pass on to others a marketing message in the form of an enthusiastic comment or story about a product or service, subtly changing the message as it goes to adapt to each individual contacted and thereby creating the potential for exponential growth in the message's exposure and influence. They key here is the level of enthusiasm and detail that accompany the message, the more, the better. Mere word of mouth pales by comparison. For example, if a new Mexican restaurant opens down on the corner of 6th and Main Street and somebody notices it and then mentions this to a friend who likes Mexican food, this is word of mouth. The restaurant owner likes this kind of free advertising. But nothing may come of it, at least not any time soon.

On the other hand, if somebody is actually lured into visiting the restaurant (by a word-of-mouth sighting or a coupon, perhaps) and has a detailed, enthusiastic story to tell when they leave ("the tacos were warm and filled with a special sour cream that was flavored with jalapeño pepper bits, the mariachi band played all my favorites, the owner came to the table and flamed the fajitas himself while explaining how they soak their flank steak in a secret marinade, the queso was made in keeping with a family recipe brought to America from the old country over 100 years ago, the atmosphere was relaxed but slightly sensual, and did I mention the complimentary valet parking for new customers?") I am perhaps exaggerating here for effect, but you get the idea. Almost anybody hearing this narrative (or reading it in an email or chat room) who is not allergic to gourmet Mexican food will want to experience this place for them self. If the food, atmosphere and service meet or exceed their expectations, they will have a slightly different and probably even better story to tell somebody else ("Somebody told me about this place but to be honest I didn't believe them but when I went there myself, well, it was even better than he said…I had the nacho burritos and…blah, blah, blah…plus our waitress sang a couple of numbers with the band…It reminded me of that Italian restaurant down on 19th Street where the waitress does an *aria* from the second act of *Cosi fan tutte* while serving the insalat…have you ever been to *that* place?…blah, blah, blah…") Now, *two* people have a great story to tell, and soon the number will be *four*, because they will each tell somebody else. Notice that the stories will probably mutate slightly and possibly become more elaborate (even exaggerated) with each re-telling, which only adds to the impact and appeal. And all of this costs the restaurant owner nothing. All he/she has to do is serve good food in a pleasant atmosphere. This is the ultimate in buzz marketing.

Make It Memorable

A product or service must be *consistently* memorable for a buzz campaign to work. Although the Mexican restaurant campaign started with a discount coupon (part of the meal, perhaps the guacamole dip, was free) it is unlikely that the memory of the occasion will place much emphasis on the cost of the meal. The food, service and overall ambience are featured in the telling and retelling of the story. The free chips and dip are a footnote. On the other hand, if the food, service and atmosphere are not really good, then the campaign will never take off. For tangible products, occasional problems are usually viewed as opportunities to show commitment to customer satisfaction. A faulty toaster can be returned for a replacement and the company selling it can actually get some positive buzz by turning a negative (faulty product) into a positive (hassle-free replacement with a smile.) But a bad service experience is usually recognized as being bad only after it is completed. Whereas a free replacement for a faulty toaster usually satisfies an angry appliance customer, recovering a customer from a lousy meal served by an indifferent staff in a cold, poorly-lighted restaurant with news TV blaring in the background is much more difficult. With most services, customer satisfaction is determined by the quality of the most recent customer experience. A string of good experiences can be reversed by just one bad experience. If the enthusiasm of the buzz message is not warranted by the actual customer experience, a reverse buzz message, complete with an elaborate "horror story," can form and spread through the same network as the original positive message. Of course, a bad experience can be overcome by a series of good ones, but in order to create a positive experience the unhappy customer has to be tempted once again to try the product or service. Getting someone to risk a second bad service experience can be difficult. It is far better to provide consistently good service, especially in situations where customer contact is frequent.

In the service business, one of the most important buzz rules is this; **customers think you are only as good as their most recent experience.** This creates pressure on the business to keep standards high. Most restaurants eventually fail not because they can't attract customers but because the food, service and atmosphere deteriorate to the point where nobody comes back (this is also why franchising has been so successful in the food industry. Franchisors keep the pressure on owners to maintain uniformly high standards.) If your service is not consistently good and you are not willing to do what it takes to keep it that way, don't encourage people to tell their friends and associates about it. Unless you have no competition, it won't work.

Be Sure It Is Networked

A buzz campaign must be networked, either face-to-face or, if it is a viral campaign, through the Internet. What does "networked" mean? Let's start by visualizing buzz marketing using simple graphic representations.

The first image below represents the direction that information flows from the top or center of a company to the prospective customers. In a traditional product promotion strategy, the marketing department develops the Features, Advantages and Benefits of the product, what marketers call the FAB formula (as in "fabulous.") Information about the product then flows from the top (the company and its Product Managers, ad agencies, etc.) downward through the media outlets, dealers and sales reps and eventually to the customers. But in a buzz campaign, most of the information that is useful or interesting about the product or service (its features, advantages and benefits) is developed by the customer and from the customer's point of view. Customers then spread their "story" by passing it sideways to another person. The difference here is a horizontal flow of information vs. a vertical flow. It can be diagramed very simply:

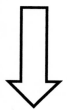

In a traditional marketing campaign, information flow is top down from the company to customers.

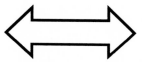

In a buzz marketing campaign, information flow is from customer to customer (peer-to-peer.)

Interviewing Tactics: A Picture Worth Thousands of Words

I have used the following question when interviewing prospective customer service workers for a position; "Draw a picture (no words) of a how you think information about our services reaches customers." Anybody who gets the horizontal alignment (more or less) of whatever they draw has an understanding of buzz marketing, even if they have never heard the term. Those who structure their answer from the top down will probably be more comfortable in a traditional marketing environment and probably also in a much larger company with a more defined hierarchical structure

The second graphic representation of buzz marketing has to do with the speed with which the message spreads through a network of peer-to-peer (person to person) communications. There is a powerful multiplier effect at work in this process, which starts with one but, if each one re-tells the story only once, increases to 128 in only 7 "generations." This graphic is self-explanatory but a helpful way to envision the extent to which a buzz message (and also a flu bug) can spread as quickly as it often does.

```
           v
           vv
          vvvv
        vvvvvvvv
      vvvvvvvvvvvvvvvv
  vvvvvvvvvvvvvvvvvvvvvvvvvvvvvvvv
vvvvvvvvvvvvvvvvvvvvvvvvvvvvvvvvvvvvvvvvvvvvvvvvvvvvvvvvvvvvvvvv
```

So why is there a need for a network? What is wrong with random message distribution to whoever happens along? A network greatly increases the credibility and accelerates the distribution of the message because it *uses existing person-to-person relationships*. For example, people who work together in an office, who "gossip" with each other every day at the water cooler (or online through email or chat) or who occasionally have coffee, lunch or dinner together as a group, are likely to exchange Mexican restaurant stories much more frequently and with more enthusiasm and detail than if they are strangers who just happened to meet on the street. Even better, their friends will believe them. Or, a good buzz network might characterize by shared hobbies or professional interests. If a product or service can identify and target a group or groups of people who might share a defined range of interests — football fans, Prius owners, college students, cat lovers — the message can be tailored to appeal to these groups and the process moves much faster. Members of a defined group speak the same language, use similar points of reference when evaluating products or services, and sometimes compete with each other to see who is "trendy" or who adapts new products the quickest. In a business-to-business (B2B) environment, a buzz program spreads faster if it targets a coherent group — construction information used by contractors, for example, or software developed for CPA's. The contractors may not know each other but they have a specific set of shared needs and interests that can cause a horizontally transmitted message to resonate. This works even better if they are members of a recognized professional trade association, which can be used as a conduit for the message. Examples might be doctors and nurses, computer programmers, antique dealers, etc. And the explosion of person-to-person social communications made possible by the digital revolution in the past few years allows viral messages (buzz using email as its primary distribution channel) to spread rapidly even among otherwise personally unconnected and geographically remote strangers who share a professional interest and who happen to view the same web pages or "meet" in chat rooms or on a blog site.

To summarize, buzz is a marketing strategy that uses people (many of whom are already customers of the business) to distribute a message about a product or service. This spreads the promotional message from person to person, like a flu virus. Internet presence is not necessary for a buzz campaign. There is no need for high-budget movie shorts or virtual pet dogs. In fact, no ad budget at all is necessary. A small business, perhaps just starting out or not yet profitable and lacking in cash, can have a tremendously successful buzz program. All that is needed is the wherewithal to *consistently* provide customers with a truly memorable experience using your product or service. Once this happens, they will do the advertising for free.

Push Marketing

Another strategy that has proven effective in promoting information services is known as "push" marketing. Push marketing has come under fire in recent years for irritating prospective customers and, in some formats, for being insufficiently cost effective as compared with overall results achieved. Consider the most prevalent forms of push marketing, which include unsolicited telemarketing, junk mail and Spam. None of these are well received by consumers. Congress has even tried to pass laws against them (colliding with a few First Amendment issues in the process but successful in curbing some of the worst abuses.) However, targeted push marketing can be very effective (and inexpensive,) especially for the ongoing promotion of information services over the Internet. It does not have to be irritating to customers and, using the Internet, can be cost effective (almost free, in fact.) Push marketing should play a role in any service business promotion strategy.

Push marketing may be defined as continuous, mostly (but not necessarily) indiscriminate distribution of information about products or services to a prospective customer whether or not the customer has asked for or has any immediate interest in the information. Worst-case examples (from the customer's perspective) include Spam, junk mail catalogs or flyers and pop-up web site ads. What they share in common is that they are not only unsolicited but also continuous, that is, pretty much impossible to stop once the sender determines that they have a valid address for a recipient. Push marketing that customers seem able to tolerate include billboards, newspaper and magazine ads, television ads, etc. These ads are fairly easy to skip or ignore, so that those being targeted can control the intensity of impact (and the irritation factor.) A best-case example of push marketing is a regular email (formerly a printed newsletter) or other form of product, company or service information that has been requested by or subscribed to by a prospective customer and that can be discontinued immediately if the recipient changes his/ her mind. In the absence of a specific request or subscription, information can be pushed to a select community of known users without causing excessive irritation. Typically this somewhat more aggressive version of push sends information about the product or service to prospective customers who, although they may not have specifically requested or subscribed to the information, nevertheless have a known or suspected interest in the product or service and so probably will not object to receiving the information. This is a calculated gamble on the part of the pusher. It may be a waste of money but there is at least some reason to believe the recipient will at least glance at the information. An example of this would be the constant stream of ad flyers from do-it-yourself homeowner supply stores

(Lowes, Home Depot, etc.) that are sent to, you guessed it, homeowners. Many are thrown away but some are not.

Even a best-case push scenario has its critics. They point out that the message is generally about price or about the product or service and the company that produces it, not about the wants and needs of the prospective customer. (By contrast, in buzz campaigns, customers participate continuously in formulating and then revising the message and so it (the message) tends to be more about customers and their needs or preferences rather than just about the product and its price.) Since distribution of the information is indiscriminant, the vast majority of recipients may have no interest in what is being sent. The waste is tremendous and can leave a vaguely negative feeling about the company in the mind of the consumer, which reduces the possibility of their becoming customers in the future if they develop a real need. Unsolicited direct mail flyers and catalogs or Spam are considered a huge success if the response rate is even a few percent of the initial distribution. The rest are wasted and end up in the landfill (or /dev/null.) The money spent by the senders on production and distribution produces no results. Recipients are forced to waste time as well, and many customers seriously resent the time and effort it takes them to manage the flow of useless information, which has to be identified and sorted from useful information, then disposed of or deleted. The typical American family receives over two hundred catalogs each year via bulk rate mail. Spam has grown so voluminous in recent years that it creates a measurable drain on Internet bandwidth, slowing global information flow and unwittingly providing a vehicle for costly software viruses. Because of the reservoir of negativity about push marketing, it can be risky for a business to engage in this kind of program. It may backfire.

But promoting information services via push marketing does not have to be negative. Useful information is, by definition, about the needs of the customer, not the company sending it out. And, unlike toasters or autos or any number of tangible products that are regularly promoted with aggressive (usually price-oriented) push strategies, information can be sampled for prospective customers, giving them the opportunity to assess the value of the service provided. If the information changes frequently, a push marketing program can capitalize on this by providing a steady stream of information updates, perhaps not of the whole body of information but at least of a certain segment or perhaps a summary update which can be reviewed quickly by the recipient. In either case, reference can be made and, if the distribution mechanism is email, active links can be provided to a web or ftp site where more complete information is available. (If the information is sold by subscription or if security is an issue, web site access may require a User ID and Password.)

The Joy of Email

In today's information intensive environment, email is the least expensive and most effective medium for a push marketing campaign. This is especially true if the product or service has information content. This view is based on a lot of good experience with my company promoting variable information services. Over time, with a steady effort to gather email addresses of customers and likely prospects through subscriptions or other voluntary customer response methods, even a small business can build a database of a few or even several thousand interested and qualified contacts. This gathering process can occur online, for example by "harvesting" email addresses of visitors to a web site, or off-line simply by asking customers for their email address when they visit or call. This database is a tremendous asset and competitive advantage (and it should be carefully secured against theft or misuse.) The addresses become the target of a consistent email push marketing program that provides free information summaries of relevant content and then (if appropriate) links the summaries to information-intensive web sites that contain much more detail. But even without a web site, an email campaign can yield excellent results. In my experience, this has proven to be an effective, almost-zero-cost strategy over a period of several years now, and technical complexity of en effective email marketing program have been simplified to point where no special expertise is needed, so it is both doable and affordable for the average business.

The ground rules for a successful (and inoffensive) email push campaign are relatively few. First of all, it is important to be aware of the Federal CAN-SPAM legislation that was passed in January 2004. Being identified as a Spammer can quickly tarnish the reputation of a business. But the requirements of CAN-SPAM are straightforward; the email should be requested (by subscription or permission,) the business or organization sending out the email must have a physical address as part of the email, and there must be a simple mechanism available to subscribers to allow them to <u>un</u>subscribe at any time (response time to an unsubscribe request is not specified by the law but it is assumed to be rapid.) These are all common-sense requirements that work to the benefit of the company sending out the emails as well as the subscribers. Compliance is highly recommended.

Above the Horizon

The second element of an email campaign is to understand how emails (and also web sites) are read and to structure the content accordingly. This is a misunderstood aspect of email and web site marketing. First of all, it may surprise you to learn much of what falls below the "horizon" of a PC or laptop screen never gets read. The "horizon" is defined as the bottom edge of the viewing screen below which an email or web site cannot be viewed unless the reader scrolls downward to reveal more text. The last line of information (or area below the last line) that is viewable without scrolling is called the horizon line. It is impossible to determine exactly where the horizon line is in any given document because this depends upon the size of the viewing screen and also the font size setting selected for the email reader or web browser on each individual viewer's PC. This is controlled by the recipient, not by the sender. Therefore, the sender can never be sure how much of the message will appear above the horizon on a standard monitor. A Personal Digital Assistant (PDA) has such a small screen that even less of a message can be viewed without scrolling.

In analog terms, the horizon may be thought of as the bottom of the page (although it really is not.) Scrolling is the equivalent of turning to the next page. All studies of digital devices show that users simply do not "turn the page" very often. They only read the top page. In digital terms, the vast majority of email readers do not scroll down through a message to read text or view graphics that are not visible when the message first appears on their screen. Therefore, whatever is important in the message has to be placed above the horizon. This greatly restricts the amount of effective information that can be placed in email but the unfortunate fact is that multiple paragraphs of text or sprawling charts will not be read by most email or web site readers. They glance at what they can see for a couple of seconds and, if it does not capture their imagination or interest, they move on, often irretrievably deleting the email in the process.

The solution to the horizon problem is straightforward; the email should contain only limited amounts of information (such as summaries or condensed data) but should then have active links to web pages or even to information below the horizon in the same email where the information can be much more complete and detailed. Think of the email as a launching pad. By clicking on active links, the reader is sent in a new direction, to a web site or ftp area for example, where more detailed information is available. There can be many active links in an email, each going to a different web or ftp site or database. Getting the reader to click on a link in an email is an important event because it is an expression of interest in acquiring more of the information that was sampled in the email. The reader is not fully "hooked" however because when he/she lands on the web site the same horizon problems exist that were encountered with the initial email. They can only see the area above the horizon. It should contain summaries or headlines with active links to other area deeper down inside the web site. But, unlike and email, the total amount of information that can be placed inside a web site is almost limitless.

Keep It Small

The size (as measured in kb) of emails and web pages is a matter of importance. There is the practical issue of the amount of information that the reader has to download into his/her local system (PC) in order to view the email or web page(s.) If the local PC is connected by broadband, download volume is not an issue. But about half of all Americans connected to the Internet at the end of 2004 were still using dial-up connections, which can be slow (usually under 28.8 kb.) Web sites that download too slowly encounter an "attention span" problem since studies (as well as practical experience) show that web site readers only allow a few seconds for a page to download. Otherwise they move on by clicking on some other link. The lesson here is that web sites cannot be designed to run well on broadband only. Until the U.S. has greater broadband coverage, webs should be designed for and tested on dialup connections to assure that all readers have a quick download experience. Making web sites download faster does not require total information to be limited but it does place limits on how much data should be placed on any one web page in the web site (a practical maximum is 250kb) including fancy graphics, photos and animated .gif files. The graphics may look snappy but significantly slow the download process. Also, graphics do not download in a logical order (from top to bottom, for example) so that a page with heavy graphic content may go through a few seconds with partial content on display, which can look strange and be disorienting to readers. They may click and move on. A rule of thumb for graphics of any kind is to keep them under 25k in size each and allocate no more than 10% of the total web page size (as measured in kb, not in actually area covered) to graphic content. But, as already noted, none of this matters if the reader is connected to broadband.

Emails also have practical size limitations that stem from other issues in regard to email design and content. For example, it is possible to design emails so that they look a lot like web pages, that is, they have a lot of color, graphics, etc. But this can cause problems. It is easy to have color text in the body of the email using almost any email composer program (like Microsoft Outlook, Eudora, etc. These are called "client" programs.) Selecting color fonts automatically converts the black-and-white text into html (hyper text markup language. Html is the universal standard computer language that enables web browsers to read information on web pages and display it on a computer screen.) Color in emails looks great and catches the eye better than plain text. But color requires html and html uses more space (as measured in kb) than plain text, so messages may move more slowly across the Internet. If information in the email is time-sensitive, this is a consideration. Much more important is the fact that many Spam filters trap and delete emails that contain html (color) because of the higher probability that color denotes advertising messages that may be undesirable (more on Spam filters below.) Extensive use of color in an email marketing campaign is a calculated risk. A certain percentage of messages will be filtered out before the intended target can read them.

Including color graphics (.gif or .jpg files, for example) in the body of an email can also cause problems. When you composing an email and including graphics, you see the actual graphic image right where you placed it for maximum visual effect. But when the email is sent, the graphic travels as an attachment to the text portion of the message. When it is received, the graphic is re-combined with the text to form the message just as it looked on your PC when you initially composed it. (In some cases, the graphic is not re-combined and appears as an attachment to the message. In this format the graphic cannot be viewed unless the recipient clicks on it separately. Because viruses also sometimes accompany emails as attachments, most readers know not to click on unidentified attachments.) When email works correctly on both ends, everything is fine except that, just as with html, many Spam filters are set to delete messages with attachments, which means your email may be completely lost before it is read. Another possibility is that the Spam filter may strip the graphic from the message allowing the text to go through but leaving an empty space where the graphic was supposed to appear. This ruins the visual effect of the email. Sending emails with graphics is a calculated risk; a certain percentage of messages will be filtered out or corrupted by partial deletion before the intended target can read them.

Did It Get There?

The problem with sending emails that never arrive or that do arrive but in a truncated or garbled form that diminishes their impact is that the sender cannot readily see these problems. If you send out 1000 emails, you assume that they all arrive. This is not the case. Some are returned to you with explanatory messages (invalid email address, recipient email box full, recipient email account temporarily suspended, etc.) Analysis by my company on returns was consistent in revealing that about 17% of all emails sent out daily were returned as undeliverable. Here is the frustrating part; the returned item list changed constantly, so that a message might get through to a recipient today but not tomorrow and then might go through every time thereafter for a month. Corrective action on returned items was often not possibly because it was not clear what the problem was or even if there was a problem. Emails might be returned with an invalid address message one day out of five. Logically, if it went through the first four times, why not the fifth time? Or perhaps it n ever got through but we were notified only one time. Moreover, we had no way of knowing which messages got filtered out as Spam because these are not sent back. They are simply trapped and deleted. No notification is sent back to the sender.

All of this points to the fact that email is not an "exact science." But there are certain simple steps that can be taken to check emails being sent. First of all, if email is really an important part of a business strategy it is worthwhile to be able to test messages through the most common email services in your area (AOL, Earthlink, Verizon, Bell South, etc.) These service providers aggressively filter messages and if yours does not get through, you can probably find out why. The best way is to start out sending a plain text message and, assuming it gets through, and then add the desired components (color text, graphics, etc.) one at a time until the message no longer gets through. If this all seems like too much work, ask customers if they are receiving the emails. They will provide helpful feedback. Insofar as corporate (Intranet) email accounts are concerned, they can be tested with the help of the intended recipient. In my experience, many employees with companies that have impenetrable email defenses maintain personal, private email accounts with one of the free services like Yahoo! or Hotmail. Customers will give you their alternate email address as a workaround if the content of the email is sufficiently interesting.

Avoiding Spam Filters

More needs to be said about Spam filters, which can destroy a push program before it ever reaches the intended targets. Spam is the worst kind of push marketing but the efforts major software companies have made to eliminate it have begun to take effect. In addition, excellent anti-Spam software is available for individuals PCs. Much of this software is free. The problem for a push program is that legitimate email is subject to the same screening that is used to identify and block Spam. For that reason, legitimate emailers need to know a bit about how Spam filters work and how they (the filters) may affect their email program.

A typical Spam filter scans (reads) email starting with the visible header (the SUBJECT: line) and contents. The filter also scans code (invisible to the reader) imbedded in or traveling with the message. Certain items, if found, trigger automatic deletion. Examples might be the words "free" or "vioxx" or the phrase "limited time offer" or the dollar sign ($.) These words and phrases are generally associated with advertising campaigns. But most Spam filters do not delete a message based just on one or two words or phrases. Instead, they "test" the message and assign "points" to many different things in the message. If the point total (called the "score") exceeds the threshold (which can be adjusted with the filtering software) the message is blocked. But if the total score falls short, then messages may get through even though they have some of the offending words or characteristics. Examples of things (besides specific words or phrases) that might earn points include: color text (html code,) graphics, active links, excessive capitalization (called "yelling",) a toll-free phone number, etc. Unfortunately these are some of the same things that may be found in a non-Spam push message. The filter cannot possibly know that that the recipient requests a message containing some or all of these characteristics i.e., it is not Spam. If the score crosses the threshold, the message is blocked.

It is absolutely necessary to test an email message format before launching a push campaign. If a format (specified color, graphics, etc.) is developed and used for every email, then it is only necessary to test the format until it works. It does not have to be checked each time. How to test? The simplest way is to send some sample emails to yourself and other recipients across a broad spectrum of email addresses (AOL, Verizon, Comcast, etc.) If the email gets through a sampling of several addressed, then the Spam score is probably low enough so that it will get through almost anywhere. Be sure to test to some company email addresses too, since these are likely being screened before being allowed to enter a corporate Intranet. Customers are usually willing to help run some tests on messages so don't hesitate to ask them. If getting customers involved is not an option, consider visiting the web sites of some Spam filter software companies. They usually have a test address available on their site so a sample message can be sent and scored. The results of any of these tests will indicate if content changes are needed (to enable the message to pass through the filters) but will not indicate what changes are the most effective. That can be determined by trial and error. In my experience, it only takes an hour or so to develop a message format and then check a couple of sample messages for filter pass-through. Because the sender never knows when a filter blocks an email, it is important to the success of the push campaign to assure that the email gets through. Remember, an average of 17% is lost each time for a variety of reasons that can be tracked (see above) If, as an early 2005 marketing study reported, another 25% are lost to filters, the impact of the campaign is diminished by 40+%.

RSS as an Email Substitute

What RSS stands for is debated among the various parties who claim to have "invented" it. The most popular definition is "Really Simple Syndication." RSS, introduced in 2004, might offer a push marketing alternative to email without some of the problems and side effects inherent in email. RSS is delivered to individual PC's by subscription only, so is not Spam. Early applications of RSS have been primarily for aggregation and distribution of news articles. A major news source, such as the LA Times or MSNBC, offers a free RSS subscription to readers who can request any news articles on selected subjects (or defined by selected keywords) that appear in the daily paper or on their web site. Moreover, the RSS aggregator also uses bots to scour other news sources for similar content and deliver them to the subscriber. The articles are not sent automatically. Instead, the subscriber is notified that new articles are available (RSS software running on your local PC has a screen that lists new "feed" (as it is known) the way your email software lists incoming emails that are available to be read.) The subscriber can then click on any items in the RSS list that look interesting enough to read. The rest can be deleted. Subscribers are notified 24/7/365 of new articles but can read at their convenience or not at all (the convenience of time-shifting.)

RSS is not limited to news sources and related articles. Blog sites (or any sites that change frequently, like the news) can also be aggregated and delivered to subscribers, with automatic notification each time the blog is updated. Advertisers have been quick to recognize the possibilities. Commercial or advocacy blog sites are established, updated frequently and then aggregated with similar sites for a direct feed into individual PC's. New features being built into RSS further enhance the use of RSS in place of email for proactive commercial messaging. To enhance blog sites with buzz/viral ad messages, blogs now offer "comment" capability, allowing visitors to respond to messages, add comments and suggestions and become an active participant in the discussion of a product or service. Most blog authoring software includes comment functionality. Another tool useful to advertisers is Trackback (also called Pingback) that automatically notifies specialized blog indexes (similar to the indexes used by search engines) when new comments are posted. Then, through an RSS feed, subscribers are notified by these index services (such as www.weblogs.com) that something new is on the blog site. They can visit, read and comment at their convenience.

One drawback of RSS is that, in order to have optimum functionality and convenience, each subscriber has to have special (but mostly free) software on his/her PC and also has to take the initiative to subscribe. Email push uses software that is preloaded onto PC's and requires no subscription activity. Special software is not mandatory. Blog summaries and links are also available from web sites that serve as aggregators. The reader does not get automatic notification of blog updates, but can access the information that is of interest, bypassing commercial messaging if desired. An example of a web site aggregator is www.lisfeeds.com a web based RSS aggregator of library blogs that is currently "scraping" (RSS lingo for constantly checking and aggregating .xml formatted blogs) nearly 200 library related blogs. Because Spam has become a major irritant, future push programs may need an alternative such as RSS to have any chance of gaining access to customers.

To summarize, the benefits of a push program using email cannot be overstated. It is relatively easy to assemble a proprietary address list of an organization's best customers. Good future prospects can be added, expanding the list well beyond the existing customer base. This list can then be provided with a regularly scheduled sampling of useful information about the services available. The *cost per contact is almost zero* so the return on effort is enormous. If it seems that I have gone into a more detail about email than with some of the other topics discussed, you are right. This may leave the impression that an email program is complicated or too difficult for the average computer user. It is not. There are hundreds of millions of people around the world who send and receive email every day. They use a small number of different programs, mostly some version of Microsoft Outlook or Outlook Express. These have a straightforward GUI interface and scores of online tutorials and Help options. Anybody can become an email expert with a couple of hours of effort. Given the enormous impact it can have, a push program should be an important priority.

11. Conclusion

Reconnect with Customers
Provide Information About Information
Go Wi-Fi
Solve the Music "Problem"
Change the Metrics
Create Digital Content
Think Sideways
Create an Information Commons

Everybody likes a good Conclusion. When I was a graduate student I always selected monographs with a Conclusion. I would read the Introduction and then check the Table of Contents for a chapter heading that seemed to best reflect the author's general thesis. I would study that chapter carefully and then tie it all together by reading the Conclusion. It was a great strategy. I could "read" two dense monographs per day and still have time to shoot some hoops with the guys before dinner. Graduate life was sweet. Of course years later I began to realize that I had only a vague notion of what the authors were trying to say and no idea at all if the conclusions were supported by substantive research. But it earned me Ph.D. in record time. (Nowadays the preferred tool is "speed reading," which means you read every word but still don't get much out of it.)

Naturally, as a writer I have developed a different perspective. I now think that skipping over most of the chapters in a book and going directly to the Conclusion is a bad idea. Unfortunately, it is still a common practice. You may be reading this right now because you fast-forwarded to save some time. It seems to me, however, that the subtleties of cultural change are difficult to appreciate from just a brief summary or conclusion. The transformation to a culture of total information is an ongoing process. The outcome is unclear. I am helping to shape it by writing this book. I email, download, blog, and do research online. I test new software and compare hardware specifications. These activities influence what I write. I illustrate many of my points by drawing upon personal digital anecdotes and lessons learned from building and then selling an information business. I think all of this makes my writing more interesting and my ideas more relevant. But if you read only the Conclusion and skip the rest, you miss some of the most useful material. If this is what you are doing, consider going back to the Table of Contents, picking something that sounds interesting and spending a few minutes with it. What does Wal-Mart have to do with libraries? What is "Ambient Information? Who are the "Millennials? You might gain some new perspectives that will influence how you use digital technology at your library. As the author, I will be gratified by your interest and effort. But, for those of you don't have the time, I have written a Conclusion. Just remember; it is not a substitute for the book itself.

The mission for libraries is to aggressively embrace the digital culture and technology of their customers. Beyond this, because digital transformation has just begun, change needs to become an ongoing activity, firmly imbedded in the library's organizational culture. What this will require is learning, planning, teamwork, risk-taking, re-thinking, training, challenging, promoting, etc., all things that are essentially "free." This can be done largely without new money. Work can start immediately. I have selected some areas where maximum progress can be made at minimal expense. Let's begin with the customers.

Reconnect with Customers

The sad fact is that if library performance over last five years were evaluated using private sector business standards, it would be judged a disaster. The customer base has defected *en masse* to competitors, and the library business has fumbled the response. If libraries were a for-profit business, the Boards of Directors would fire the CEO's. Turnaround consultants would be brought in to engineer a complete shakeup of the remaining management and staff see what, if anything could be salvaged. If the prognosis was poor, the whole business might be put on the block and sold by the parent company. Worse, it could simply be closed and any necessary services outsourced. (Is it any wonder why non-profits are frightened by private enterprise?) Prior to the mid-90's, libraries held a monopoly on the services they provided (information) and the clientele they served (mostly students.) But this monopoly is gone and, like any service business losing market share, libraries have to rethink their strategies and retool their offerings. Failing that, they will be squeezed out of the market altogether. From a private sector perspective, the CEO's need to start thinking of themselves as "turnaround specialists."

The first priority for any struggling service business is to reconnect with its customers. Transitioning from service monopoly to service competition will necessitate changes in both content and delivery. Library competitors are not invincible. They have major problems with their services (see Chapters 4 and 5.) They also have hidden strategies and agendas that are not necessarily in the best interests of their customers. These weaknesses can be exploited. One of the underlying themes in this book is that libraries can offer competitive, even superior information services if they are structured properly and delivered consistently. (I evaluate the head-to-head confrontation between libraries and their competitors in Chapter 3.) The competition will be good for libraries. It will stimulate greater effort and result in better service. Best of all, the libraries' customers will be the ultimate winners.

Students and Faculty are the library's customers. Treating them as such must become imbedded in the library's service culture from the reference desk all the way to the back of the Back Office (see Chapter 8.) This is not pandering. It is central to the mission. Here are some places to start.

1. Change spaces (see the section entitled "Changing Spaces" in Chapter 9.) A key difference between libraries the Internet is that libraries can provide a user-friendly, "retail" environment. This can be a major advantage. Physical reorganization of facilities, even if it does not involve installation of new equipment, can exploit this natural advantage and send a powerful signal to staff and customers alike that change is coming. Continuous change sends a continuous signal, so not everything has to be done at once. But it is important to start immediately.

2. Survey the customer base. Do it continuously. (See Chapter 7, "Digital Students.") The goal of marketing surveys is to identify customer needs, expectations and level of satisfaction with current services. This can be accomplished with online surveys, over-the-counter surveys, focus groups, advisory boards and any number of other techniques that seem appropriate for a specific library. As I discuss in Chapter 7, the impact of customer feedback can be powerful even if the feedback itself seems useless. If 100 customers give bad advice about how to solve a problem but they are all talking about the same problem, the survey is well worth whatever effort it takes. The mere act of asking for the input can have a positive impact on customers through the signals it sends. Also, top management should make a point of talking with (informally surveying) customers. As CEO of my service business, I had a personal goal of speaking to at least one customer each day to find out "how we were doing." I was thrilled with some of the responses and disheartened by others, but I gained invaluable insight into what our customers were thinking. With this information, I could

correct problems and identify new programs that customers wanted. A library organization is not so big that top management cannot be involved with their customers. The feedback will be well worth the time and effort.

3. Hire people-oriented employees that have both the aptitude and attitude to bring information services to others. As I discuss in various spots throughout the book, a customer's evaluation of service quality is based on *perception*. The right human interface is a vital part of creating the perception of excellent, above-and-beyond-the-call-of-duty service, which should be the standard set by any service business that hopes to succeed. As I discuss in Chapter 9, although it is still not easy, it is easier than it has ever been to hire the kind of digitally capable service personnel who can connect with customers. Take advantage of turnover and any new positions that open up to staff properly for superior customer service. Reassign existing staff as necessary. It will pay dividends in customer satisfaction and loyalty.

4. Promote library services. Do it continuously. Back in the days of monopoly, libraries may not have felt the need to promote themselves. But times have changed (as they did for doctors and lawyers) and a competitive environment demands that a service provider must be aggressive in getting the message out to the customer base in order to be successful. I devote a whole Chapter to "Promoting Service" because it is of vital importance. Using buzz and viral tactics can also be a lot of fun, and using push strategies is pretty much free so use can be unlimited. Today's millennials are accustomed to promotional activities targeted at them that are properly attuned to their culture, so that much like the act of surveying, the effort of promotion itself sends a message that is important. Promotion will pay dividends in drawing attention to library services and raising customer awareness of why and how the library should be the preferred source for information.

Provide Information About Information

The business model developed by Internet search engines is totally dependent upon the sale of paid advertising and advocacy web site links. The profit potential is enormous and this trend is sure to continue. Increasingly, as I argue in Chapter 5, commercial information is not identified as such but instead is intermingled with the non-commercial information produced by customer searches. The growing popularity of news bots (automated search software) and RSS (Really Simple Syndication) is creating even more opportunities to infuse commercial and advocacy messages into news articles and other forms of traditionally non-commercial information. Along with penetration of the formerly ad-free "blogosphere," news-related advertising is expected to be the next highly profitable Internet frontier.

What is troubling is that all studies, but especially data gathered by the Pew "Internet and American Life Project," show that search engine users are only dimly aware of the commercialization of the Internet. The typical user rates search results as a "fair and unbiased" source of information. Based on empirical search studies, they are wrong. Ironically, part of the credibility of search engines, a credibility that is carefully nurtured by the search engines companies themselves, stems from uncritical endorsements by librarians, many of whom routinely recommend search engine web sites as a supplement to more traditional resources. Inadequate effort is being made to explain differences between the two. Given what Pew calls the naïveté of the vast majority of search engine users, this is irresponsible.

Libraries should initiate a training program (give it an exciting name that includes the word "initiative," something like Academic Research Initiative—ARI for short) to make search engine users more aware of the benefits but also the shortcomings of online search. (Other information sources should be studied and critiqued as well.) This is an important responsibility for the academic community and an ideal mission for libraries. The goal of this project should be to educate students and the community at large about digital content and the differences between it and more traditional content from libraries, textbooks, periodicals, monographs, etc. Libraries are in a unique position to advocate careful user evaluation of information and to help users understand the differences between information and knowledge. This is an important part of helping students develop their capacity for critical thinking (which depends upon knowledge gained from a broad spectrum of reliable information.) To this end, libraries can serve as a catalyst, by

1. Creating (and continuously updating) online tutorials to help library customers learn about the latest search engine strategies, tactics, strengths and weaknesses. This would necessitate moving well beyond the current practice of merely listing URL's for search portals, databases, etc., which is typical of too many library web sites.

2. Promoting ongoing discussion of search engine content and related issue by means of interdisciplinary seminars, a guest lecture series, etc.

3. Researching search engine evaluation web sites and professional organizations such as the American Library Association (ALA) for insights into search engine content and how to best use it in an academic application.

4. Evaluating and distributing information about search services and techniques provided by the search engines themselves.

Libraries can draw upon interested administrators, Faculty members and community resources (experts in advertising, law, ethics, epistemology, economics, information technology, information management and library science) as guest speakers, bloggers or as guests or moderators in live chat sessions. Libraries can also use buzz marketing to create awareness and encourage debate.

Libraries have always been a trusted source of information; now they need to become a trusted source of *information about information*, especially as it pertains to the Internet search phenomenon. Students need this. This is a realistic, achievable goal for all libraries and librarians. It should be written into the mission statement of every library.

Go Wi-Fi

Wi-Fi is rapidly becoming the connection of choice for millennials. I estimate in the section in Chapter 6 that Wi-Fi will be a reality throughout the U.S. in 24 to 36 months. The demand for ambient information has already begun and it will grow as ambient access grows. The appeal of ambient access is that it facilitates "space shifting," the ability to compute anywhere. (The appeal of its counterpart, ambient information, is that it facilitates "time shifting," the ability to access the resources necessary to compute any time. I discuss this concept in more detail in Chapter 6.) Ambient access and ambient information are two of the holy grails of computing for today's highly mobile, 24/7/365 population.

Libraries should make Wi-Fi available throughout all facilities. In doing so they will send a strong signal to students that the library understands and supports the culture of total information. All libraries, whether university or community-based, should become Wi-Fi hotspots. Wi-Fi is one of the least expensive new technologies to install and maintain, much cheaper than hard-wired connections, and it is easily upgraded as improved protocols (Wi-Max, for example) or new equipment become available. Students and other customers should be encouraged to bring their laptops and other wireless devices to the library. Work space-laptop workstations with AC power strips-should be made available even if it requires reallocating floor space, retiring older study desks and, over time, unplugging some networked PC's to make room. A Wi-Fi "expert" (someone from the library staff or perhaps a student or community volunteer) should be available to assist anyone having problems (logging on for the first time, for example.) An on-line tutorial could substitute for staff.

Wireless is not limited to computing and so there should also be a cell phone area where use of these or other wireless devices such as PDA's, is not restricted due to noise or electronic or structural interference. This will become a social place, so soundproofing or physical isolation has to be considered. But it will be worth the effort and relatively small expense. For millennials, access to friends via cell phone is critically important to their life. The unrestricted availability of Wi-Fi should be promoted with as much buzz as the campus and community will tolerate. (I discuss the power of buzz marketing in greater depth in Chapter 10, "Promoting Service.") By launching a major Wi-Fi initiative, the library will be signaling that it is expanding digital services. Their customers will be quick to take note.

Solve the Music "Problem"

A lot of Chapter 6 is devoted to pop music, and I encourage librarians to become more engaged with current issues regarding online music and video. The reason is that music today is about technology as well as entertainment. (It is also about legal matters, business models and a whole host of other topics that were meaningless to earlier generations of music fans.) Today's millennials and thumbs have enjoyed a broad expansion of musical genres and entertainment personalities, all of which are readily available anytime, anywhere thanks to digital technology. These are the library's customers. A reluctance to embrace such an important element of student culture only serves to increase the disconnect libraries have experienced. It is time for libraries to face the music.

The overall library strategy should be:
1. Take a high-profile position against use of library equipment and networks for "illegal" P2P (or other) music and video downloads <u>until</u> the intellectual policy struggles are resolved by the courts,
2. Partner with one of the currently "legal" music download services to offer library customers online access to as much music as possible, preferably at a steep discount
3. Promote all of these activities with a high-impact buzz campaign.

Let's look at each of these individually.

First libraries should take steps to assure that their equipment and network cannot be used for "illegal" music downloads. This matter will be resolved in the courts over the next half decade, but in the meantime the best strategy is to *sidestep it*. The library does not have to agree with the RIAA that downloading is "illegal." But the first step in a good defense against litigation is a good defense (ask the appropriate legal counsel.) In Chapter 6 I discuss five possible defensive IT and administrative options that are available (there are more, but they are also much more costly.) One or more should be deployed. The library's position should be that until the controversies over downloading are resolved, the library takes a conservative attitude on the matter but also reserves the right to change that position as the courts clarify matters. As the library educates students about the legal issues involved, students will understand the library's need to be cautious. Librarians, in turn, need to understand student angst over these issues and show some empathy (see "Heard in the Dorm" in Chapter 6.)

Next, the library should go on the offense by contacting any and all of the "legal" music download sites, picking the one that seems most appropriate, and creating a high-profile "partnership" to promote the availability of online music for the library's customers. (In Chapter 6 I list a handful of universities that already started testing this approach. Check with some of them to see how it is going.) Typically these download sites offer sharp price reductions for "tethered" music—music that is available anytime, anywhere but only as streaming audio. Tethering allows for unlimited access and playback though a wide variety of compatible devices but prevents permanent downloading, saving, copying and redistribution. Full "legal" downloading is available on a per song basis but only for an additional fee. Competition among the many fledgling music download sites has resulted in aggressive marketing efforts and good discounts. *Napster To Go*, for example, claims to have one *million* songs in their collection and charges $14.95 per month for unlimited tethered access. Napster charges $.99 for individual digital copies that can be downloaded. (These are list prices. Actual prices are probably a lot lower for a bulk subscription. Free trials are common. I am not recommending Napster, only mentioning them as an example. Smaller download companies might be less expensive and more cooperative.) The library would establish the service but individual students could be asked to pay for his/her actual usage using at the library's discounted rates.

Lastly, the library should use the music solution to create a high-impact promotional program. Since the music is digital, the promo should be digital, but word-of-mouth starting at the front desk would be enough to get the buzz started. Perhaps a few free subscriptions could be given away as part of the promotion (the download service selected would probably contribute a few at no cost.). Student customers could register for the giveaway online or by filling out a form each time they actually visited the library, whichever seemed most appropriate. A "drawing" could be held every 30 days. This would serve to keep the momentum going and give new customers an opportunity to win.

Hardcore downloaders might scoff at this overall solution but the library would be sending the signal to its customers that it understands that music is important and that the library is willing to make the effort and flex whatever muscle it has (in terms of buying or computing power, for example) to make "legal" music available. The approach suggested would deflect potential litigation problems and clear the way for other uses of P2P information transfer which right now is underutilized due to its close identification with "illegal" music downloading (see Chapter 6.) It would also send the right signal to students and thumbs by helping them secure low-cost access to music, and cost the library almost nothing. Work on this project should start immediately.

Change the Metrics

As the services provided by libraries change, the statistical measurement of library activity should also change. Otherwise, the utilization of library services may be misinterpreted and underreported. Libraries have become accustomed to quantifying activities some of which now appear to be in decline as use of digital technology becomes more widespread. But digital technology is growing in part because it allows both space and time shifting, neither of which draws people into the library. For example, increased use of online information may reduce the number of visitors who come to the library to borrow books or do research. (The occasional flap over "library hours" is an example of a time shifting clash between a library and its customers. The library cannot win this battle because even if they stay open 24/7/365, which some are starting to do, users will still want to space shift by accessing services from home or work.) Statistics may indicate that a library is servicing fewer customers when in fact it may be servicing more customers but in other ways using digital technology. If the right digital services are available, customers will use the library online instead of in person. The right metrics are necessary to accurately track this change.

Reference desk use may also be changing. As the availability of online information resources and email and live chat reference desk support increases, it may appear that use of reference desk services are declining when in fact they are not. Also, as I discuss in Chapter 3 in the section "Academic Search the Google Way," new indices are needed not just to measure a library's activity levels but also to measure customer satisfaction with library services in areas such as reference desk response accuracy and thoroughness, where there is anecdotal evidence that libraries are doing an excellent job compared with search engines.

Finally, new library activities such as the "Academic Research Initiative" that I have suggested will need to be quantified. As library resources are redirected toward projects of this type, library activities will shift. Quantifying growth areas will serve as a statistical counterbalance to areas that are in decline. As the library's use of its resources changes, so too should the measurement of return on those resources.

Create Digital Content

I have recommended sidestepping programs that might put libraries in conflict with copyright holders until intellectual policy issues are clarified by the courts. This will take time, perhaps years. Libraries have understandably shied away from any action that might run afoul of copyright holders. Some universities have been sued by the RIAA for allegedly allowing their networks to be used for "illegal" music file download activities (see "Suing John Doe" in Chapter 6.) Fear is an effective deterrent and as a result of lawsuits both actual and threatened, libraries have curtailed many activities because of the problem with music files. These include a reluctance to start converting library book collections into digital content and making them available online to library customers.

Google, on the other hand, in conjunction with a small number of large libraries, has decided to begin a scanning program that will convert 20 *million* books into digital format so that they can be made available to library patrons and web searchers over the Internet. For works in the public domain, the full text will be made available. For copyrighted works, only a small portion will be available for online viewing. This program is bold. It is unclear whether authors or publishers will challenge the right to scan portions of copyrighted works. Google is interested in pushing the envelope, however, since their long-term goal to make all information (including all books) available over the Internet and since they probably have the cash and the in-house legal staff to fight any challenges. Individual libraries cannot take those risks.

Avoidance of conflict may be prudent, but at what point does the fear of copyright litigation deter libraries from undertaking projects that are both legal and of value to their customers? As an example, a large number of older works, estimated to total as much as 30% (or even more) of the content of a typical library, have entered the public domain because the author's copyright has expired. U.S. copyright protection for works published prior to 1978 required that for a work to be copyrighted forms had to be filed and fees paid and then the copyright had to be periodically renewed in order for the book to receive copyright protection. It is estimated that half of all books published prior to 1978 were never properly copyrighted to begin with and, of those that were 90% were never properly renewed when the initial copyright expired. (Today, due to changes in the law, copyright is automatically assigned without need to register or even to affix the copyright logo.)

What all of this means is that most libraries contain large numbers of books that are not copyrighted. These can be scanned to digital format and made available in their entirety without legal consequence. Preparations should be made to start this process. It will take time. First of all, it is not always easy to determine if a book is (or was) copyrighted properly, especially if it is no longer in print. There is no central registry of current active copyrights and publishers have been reluctant to provide information, perhaps because they feel that generalized uncertainty contributes to intimidation. Also, while scanning may seem easy, high volume scanning and indexing (a necessary part of the scanning process) is essentially a manufacturing process that requires good organization and rigorous quality control of the finished product. There is a learning curve and standards have to be adopted. But the process could start with internal library documents such as forms, lists, procedures, etc. for which copyright is not an issue. These could be scanned and either made available through the library's web portal or distributed by email in response to customer inquiries. Larger scanning projects might exceed a library's in-house capabilities and could be outsourced either to companies that specialize in scanning and indexing or to local digitally capable quick copy shops, most of which have the necessary scanning expertise. A high-volume long term scanning contract would command deep discounts from print shops that might make it affordable even on tight library budgets. The library then might consider freeing up prime library shelf space for other uses by removing the hard copy versions of scanned books from library shelves. Digital files would be stored on servers and hard copy would be warehoused. Space savings would help offset scanning costs.

Most libraries have been overly careful not to stir up copyright issues. It is time, however, to begin doing what is perfectly legal (and ethical) whether publishers like it on not.

Think Sideways

Rigid, hierarchical management structures have rarely excelled at delivering superior customer service when human interface is critical to the transaction. If you are withdrawing money from an ATM, you want precise execution of commands with predictable results. There are no gray areas; you either get the right amount of money or you don't. A programmed machine can usually do the job satisfactorily. But if the ATM fails and you call the bank's customer service department for assistance, you want to speak with someone who has the authority to use the bank's full resources to solve your problem. You don't want to hear about process, procedures, rules and regulations and that they will have to discuss it with their manager and get back to you. You want results immediately. And you want to be made to feel good about it while it is happening. An ATM machine cannot do this. It takes a dedicated human being.

Small businesses frequently outperform large corporations in delivering services (if big businesses provided superior service, we would all be sending our dry cleaning to General Motors.) Why is this? I think it is because in a small business, the owner him/herself (with the assistance of a small group of employees) is directly involved with the customer. There is little hierarchy. Communication and decision-making are owner-to-customer, not owner-to-manager-to-supervisor-to-customer service representative. Larger corporations have tried to mimic the immediacy of direct communication by means of management strategies such as "empowerment" which was popular in the late '80's and early '90's. But the organizational complexity of large businesses and especially the need to make decisions that are determined primarily by budgetary considerations make it difficult for the employees who actually work with the customers to function like owners or top management. By comparison, the small business owner or employee is automatically empowered due to the size of the business and the fact that, with a small number of employees relative to the number of customers, everybody who works there is likely to have constant customer contact. If they are properly trained and motivated this can be a powerful factor in making the business succeed.

The idea that the best customer service occurs without hierarchy is what I call "thinking sideways." Structurally it means that **information and services flow directly between people and do not have to be processed through some central point in an organizational structure or communications network.** I discuss this concept in more detail in Chapters 9 and 10. In the pre-digital era this business strategy was successful in the form of franchises, an organizational structure in which businesses, even those with huge market potential, were broken down into small, independent units that were guided by but not controlled by a much larger organization. It is no surprise that franchising prospers in the service sector more than any other. The strength of a franchise is the freedom its owner has to communicate directly (sideways) with customers thereby exerting immediate influence on the level of customer satisfaction. Franchising remains one of the fastest growing and most successful business strategies in the U.S.

As digital technology has evolved, sideways thinking has taken on added dimensions. The universal spread of GUI interface, which I discuss in detail in Chapter 9 on "Training" has called into question the need, effectiveness and expense of top down employee training. Sideways (peer-to-peer) software training is faster, less expensive and as effective as more traditional top-down instructional methods. I likewise suggest that traditional hiring criteria for service workers be changed to increase the number of employees who have a predilection for "thinking sideways." Sideways thinking is also the core idea behind buzz/viral marketing (see Chapter 10 on "Promoting Service.") Allowing the customer to not only carry but also develop the advertising message, one of the purest forms of sideways communications, revolutionizes a service organization's ability to reach potential users. Online auctions such as *Google Answers,* which I discuss in Chapter 3, also utilize sideways thinking. In a top-down business environment prices are set by corporate managers and then negotiated between customers and lower-level sales or customer service representatives. In an auction, price is determined by direct customer vs. customer bidding (sideways communication.) The seller of the item is not involved. Auctions have been one of the fastest growing online activities over the past few years not just because they are fun but also because they have proven to be highly efficient (in economic terms) and therefore satisfying to buyers and sellers alike. And finally, don't forget peer-to-peer file exchange, which I discuss in depth in Chapter 6 "Ambient Information" and again below in the section on encouraging a local "information commons." P2P is a communications architecture that allows pure sideways digital information exchange. This is exactly the problem the entertainment industry faces, that is, a breakdown of centralized control of entertainment information distribution, the lynchpin of their business model for the past many decades.

Successful service businesses have learned that service improves when hierarchies are not directly involved with customer transactions. And digital technology has created new opportunities for sideways (peer-to-peer) communications within an organization as well. This is something that libraries, in need of winning back customers with superior service and controlling expenses through superior efficiency, should heed. Sometimes the best moves are sideways.

Encourage an Information Commons

Although the direction the Internet will take in upcoming years is hard to predict, if the past three years are any indication the emergence of new technologies will make the it much easier for information based organizations such as libraries to create and maintain an "information commons" on their campus or in their community. Among these technologies are P2P file transfer, weblog hosting and authoring tools, RSS information delivery, and audible information distribution, the most interesting being podcasting. Each of these has advantages and disadvantages for information exchange but they all share some common themes, including the potential for locally created, unfiltered distribution of information using network architectures that allow direct and in some cases two-way communication between individuals or between and among individual members of an group. Although some uses of these channels will no doubt be heavily influenced by centralized commercial and advocacy interests (this has already started,) they nonetheless offer potential alternatives to the commercially-driven information sources that have come to dominate the Internet in the past few years. As the preeminent source for information on campuses and within communities, libraries should look for opportunities to encourage use of communications technologies and promote the idea of an information commons

What is an information commons? It is the cyber world counterpart to a physical world park, town square or street corner, a place accessible to the public for (mostly) unrestricted free-form activities. Coffee historians point out that cafes serve this purpose in some cultures and that their growing reputation as an information forum (as Wi-Fi hotspots, for example) helps explain their popularity in the U.S. (see the section on Starbucks in Chapter 2, above.) Bookstores (also popular Wi-Fi hotspots) serve the same purpose in many communities. Commons areas are everywhere. Most small towns have a downtown park, often inside a square of public streets or walkways. Larger cities have many parks and recreational areas scattered about. On a larger scale, county, state and federal governments maintain public areas, open for leisure use by all citizens. Sometimes a token access fee is required, but these places are mostly free. On any given day commons areas are the site of softball tournaments, chess matches, political rallies, family picnics, touch football games, swimming, July 4th Bar-B-Q's and Frisbee tosses. When the carnival comes to town, it usually sets up in a commons, even if it only a parking lot. Of course there is a down side to free commons; they are sometimes festooned with dog poop and they can become a haven for obnoxiously loud people, drug dealers, muggers and worse.

Political philosophers and pundits from John Dewey to Anatoly Sharansky have linked public space (physical commons) with the "public sphere," (a free-flowing information commons) often cited a prerequisite for successful democracy. Writing in the early 1920's Dewey called for an expansion of the public sphere in the "local community." This should include, he said, a broader cross-section of the public in order to counter the growing influence of elites who, according to Dewey, have a vested interest in the centralized and filtered distribution of information. More recently Sharansky, a former Soviet dissident, has argued that the litmus test for democracy is the ability of citizens to go to a public square (or sphere) and say whatever they want to say without fear of reprisal. So an information commons means more to the body politic than just a simple way to swap digital information. Its mere existence fosters and helps define democracy.

1. It is difficult to say which information exchange technologies offer the greatest promise. Peer-to-peer (P2P) file transfer could be a powerful factor in the future of online communications. In just a short period of time, tens of millions of users have downloaded and installed P2P software and file transfer activity is measured in the millions of files *per day*. P2P represents a major change in computing architecture, a remarkable transformation of information exchange from vertical (top-down) to horizontal (sideways.). Whereas web browsing, telnet, file transfer protocol (FTP) and other communications protocols require information processing through a centralized point (a server) P2P allows direct PC to PC communication between individuals. Although the Recording Industry Association of America (RIAA) has attacked P2P because of music downloads, even it has begun using P2P to gather marketing data for the record industry. The Motion Picture Association of America (MPAA) has been more careful to limit its criticism to what it alleges is illegal use of P2P, not to P2P itself (see Chapter 6 and especially the section entitled "Suing John Doe.") Litigation against P2P has slowed its acceptance by businesses and other organizations, but P2P will likely play a major role in undisputedly legal information exchanges of all kinds in upcoming years. It is inexpensive (free in many cases) and easy

to use and libraries could easily create and manage P2P information sharing folders that could be accessed by users worldwide.

2. Weblogs (blogs,) discussed in Chapter 5, have soared in popularity as authoring tools have become more sophisticated and easier to use. According to Merriam-Webster Inc. the word "blog," defined as "a Web site that contains an online personal journal with reflections, comments and often hyperlinks," was the number one word looked-up on its Internet sites in 2004. A Pew survey indicated that 27% of all Americans online in 2004 read blogs and 7% wrote blogs. The Merriam-Webster definition may already be changing. New blog technology allows two-way communication via the posting of comments by blog readers who instantly turns a blog into a forum for exchange rather than just a one-way rant or rumination, which was characteristic of most early blogs. Video blogs—called "vlogs"—allow posting of digital film clips together with text. Blog content originating from cellular phones-called "moblogs" as in "mobile blogs"- have made blogging wireless by allowing people to post video and photos taken with camera phones to a blog or to add an audio posting directly from the phone. Weblog hosting is also available from a variety of online sources (www.blogger.com, for example) most at no-charge if you are willing to tolerate some identified ads placed on the landing page. As with P2P technology, blogging has its detractors, some from within the library community. Michael Gorman, a professional librarian for 40 years and author of an interesting book on librarianship in the 21st century and currently President of the American Library Association, has lashed out both at blogging's undeniably rough edges and the "fanatical belief" in the technology that grips blog advocates;

"[The] Blog People (or their subclass who are interested in computers and the glorification of information) have a fanatical belief in the transforming power of digitization and a consequent horror of, and contempt for, heretics who do not share that belief...Given the quality of the writing in the blogs I have seen, I doubt that many of the Blog People are in the habit of sustained reading of complex texts. It is entirely possible that their intellectual needs are met by an accumulation of random facts and paragraphs.'"

Others within the library community are enthusiastic about blogs. As an example, Stephen Cohen, a librarian and an early adapter of blogging maintains a blog site www.librarystuff.net that educates and advises on the science and art of blogging. In the debate between Gorman and the Cohens of the blogosphere, the shift to a culture of total information makes blogging an odds-on favorite.

3. RSS (Really Simple Syndication) discussed in Chapter 10, enables information flow to interested readers from a wide variety of sources ranging from newspaper sites to blogs. This makes it much more than just an alternative to email. RSS gets easier to use with each passing day, and is now being linked not only to information intensive web pages but also to blogs. This makes blogs far more accessible to a broader array of readers. Automatic notification (by free subscription) moves information and conversations along very quickly, if appropriate. Web based aggregation eliminates the need for special software. From an information distribution perspective, RSS lowers the barriers to entry for anyone wanting to get a message out to a larger group.

4. Podcasting is a new form of information distribution that takes advantage of the growing number of iPods being used by time-and space-shifting millennials. There reportedly are now 11 *million* iPods and an unknown number of iPod clones. They all are capable of storing and playing back MP3 digital audio files, which can be downloaded from the Internet or loaded into the iPod from a PC. Listeners can then play the MP3 while exercising, relaxing, working, etc. Creating MP3 audio files is now simple enough that it can be done at home with little equipment other than a PC and a microphone. Podcasts are currently similar to early public access TV as immortalized in the Saturday Night Live *Wayne's World* skit; professional quality is not as important as the message. Podcast titles — *Say Yum* (a food show) *Why Fish?* (a weekly show from North Dakota) *Grape Radio* (about wine) — reveal that early adapters have made this a highly specialized and personalized, much like the early days of blogs. And, like many blogs, podcasts have small listenership, although former MTV host Adam Curry claims to have several thousand listeners to his podcast called *Daily Source Code*. Bigger players are starting to get involved. Public radio station WGBH in Boston is podcasting its *Morning Stories* show, which had only 30 downloads in its first week but 57,000 downloads during the month of December, 2004. The popular technology web site ZDNet launched its podcasting in January of 2005 with some highly technical audio articles. Podcasting already has its weekly top 50 list available at www.podcastalley.com (motto: Free the Airwaves.) Listeners are notified of new podcasts by means of RSS feed directly to their PC. Software is available to automatically download MP3's onto a local hard drive as they become available. They can then be heard directly from the PC or loaded into a portable device (an iPod) for later use.

What content might find its way into a digital information commons? Like the evolution of the technology itself, it is difficult to say. The size of the community could be open and global or could be limited by firewalls just to "insiders" so we would expect content to reflect the interests of a local campus or community of students and residents. A common theme at a liberal arts campus might be creativity, whereas an engineering curriculum might reflect more technical interests. Students could share portfolios of art use P2P or short stories or poems from creative writing classes via blogs. Readers, notified by an RSS feed, could comment back about style and content. Professors might use P2P to publish groups of student term papers or exam essays for others to read, thereby sharing research and insights across a broad spectrum of students and classes. Lectures (or at least the outlines) could be published as blogs, thereby inviting student questions or commentary that would be available to everyone else in the class. Community associations could post the minutes of meetings, using RSS feeds to notify constituents of changes or updates to community budgets, master plans or zoning regulations. This could facilitate letting the public in on planning and policy decisions, something that is normally not done in part due to constraints on available information about the issues. University administrations could do the same. Libraries could distribute customer tutorials and establish user blog sites to invite comments and suggestions. Libraries might also consider podcasts of special lectures or performances for those unable to attend in person. (Podcast articles and books are already being used to augment programs for the blind.) Religious groups are also starting to podcast sermons, which can be played (and replayed) by parishioners as time allows. Advertising would almost certainly make its presence felt, though perhaps not from global companies. Campus and community entrepreneurs could hawk their wares (perhaps as sponsors of a podcast) and business majors might want to use the commons to promote or test market their latest projects. The Young Republicans (or Democrats) could blog white papers blasting the local mayor's latest budget schemes. Local pajamahedeen could vlog or moblog everything from local sports events to political rallies using nothing more than their wireless camera phones. Instant notification to readers via RSS feeds would make this kind of "reporting" available in almost real time, IT experts could propose solutions to local wireless access problems. Geography might also play a role. The local or campus police could publish updated diagrams showing parking restriction schedules in areas under construction or near classroom buildings, thereby aiding commuters. Student feedback via comments to blogs could help shape many policies that affect campus life. City and county governments could provide maps showing voting districts in local neighborhoods. All (or none) of the above might find their way into an information commons, plus a lot more that we cannot know.

While much of this type of information could also be made available through web sites, in my view the Internet, though only a decade old, is already becoming too diverse and commercially intrusive to effectively serve local community needs or interests. As the influence of advertising and advocacy on web information grows, web sites and now even some commercially subsidized blog sites seem more interested in gathering information about readers than dispensing information about issues that matter to a local community. Web content is centrally developed and deployed so as to optimize the tailoring of the message and maximize the use of data gathered. Even completely passive web site visitors leave a trail of information behind (for example, what URL they came from and where they went after leaving) and take away implanted cookies or other identity markers. In the physical world one can read bulletin boards and even attend political events without actually getting involved. In the world of the global Internet people like this are derisively labeled "lurkers" and sophisticated technology is deployed to find out who they are, where they came from, where they are going and what interests them. But not everyone wants to be in somebody else's database, and an information commons could offer an alternative. In a commons using P2P or RSS, nothing would be sent to anybody unsolicited and nobody would have to read anything they didn't want to read. Like a physical bulletin board, content could be quickly surveyed and anything that looked good downloaded and read. The rest could be ignored. People would be free to participate or not, as they saw fit.

The idea of a digital information commons is not radically new but neither does it have proven value, at least not yet. There are risks. It could fail and its proponents become a laughing stock; there would be recriminations. It could be risky content-wise; there would be community outrage. It is a bit more complex technically and could be a drain on the technology budget; opponents would criticize the waste of limited resources. We cannot be sure if democracy would be enhanced in an information commons as Dewey and Sharansky (and many others) suggest. It might, but probably not without a lot of organization and forbearance. There could be unintended consequences.

Despite the uncertainties, I think promoting an information commons would be an ideal project for libraries, the preeminent local campus and community information centers. The library could help the local on-line community define information parameters, establish information categories and promote use of the commons to all library customers and community members. This would be a substantial undertaking and would require coordinating the interests and efforts of Faculty, campus and community activists, administrators, library staff, student groups, etc. Like a physical commons, an information commons might also attract riff raff, which would severely test the openness and commitment to democratic process, just as it does in the physical world. The library would have to learn when to intervene and when to stay out of the way and let activities in the commons run their course. But a digital information commons would be an exciting experiment both in information distribution and democracy and, if successful, an immense contribution to the community. I hope this project becomes part of the agenda for libraries all across the U.S.

* *

Libraries are a great institution, a human treasure. They find themselves at a crossroads, hard-pressed by declining budgets, Internet search technology and cultural changes that threaten their traditional role in society. Thoughtful, forward thinking people need to rekindle their interest in libraries and in the vital role they play in an informed, prosperous and democratic society. This is important to our future. Let's get started.

I'll see you at the library.

Index

A9 (Amazon), viii, 33, 41, 42, 43, 68
Academic Research Initiative, 165, 170
Adelphia, 76
administrators, 5, 8, 9, 14, 70, 98, 100, 112, 114, 166, 179
advertising, 12, 29, 34, 38, 44, 45, 46, 52, 53, 55, 56, 57, 58, 59, 60, 61, 62, 65, 66, 67, 69, 70, 71, 72, 91, 105, 128, 146, 151, 156, 158, 164, 166, 173, 179
advocacy, 47, 56, 59, 65, 66, 67, 159, 164, 174, 179
aggregators, 160
AI, 40, 50
ALA, 5, 165
algorithm, 37, 63
Amazon, viii, 33, 41, 42, 43, 45, 68
ambient, 16, 71, 73, 74, 86, 166
Ambient Information, ix, 13, 46, 71, 162, 173
Ambient Orb, 73
American Historical Association, 8
American Library Association, 5, 165, 176
analog, 10, 12, 13, 37, 45, 72, 84, 85, 100, 115, 116, 117, 118, 124, 126, 135, 137, 139, 140, 154
AOL, 33, 45, 51, 157, 158
Apple, x, 27, 63, 85, 89, 96, 127, 128, 129, 130, 132, 139
art, 7, 89, 176, 178
artificial intelligence, 40, 50
Ask Jeeves, 33, 34, 45, 48, 55
Aspell Spell Helper, 40
Association of Research Libraries, 37
AT&T, 76
Atlanta, 74, 78, 81, 95, 96, 124, 129
auctions, 173
audio, 21, 71, 72, 97, 102, 168, 176, 177
Autozone, 87

back office, 116, 118
Barry, Dave, 25
behavioral targeting, 53
Bell Telephone, 76
Berkeley, 8
Betamax, 92, 96
BigChampagne, 95, 96
Binghamton University, x, 2, 5, 125
BitTorrent, 91, 95
blinkx, 33, 49, 50
Blog People, 176
bloggers, 62, 70, 100, 166
blogging, 9, 176
blogosphere, 61, 165, 176
blogs, 61, 62, 89, 144, 159, 160, 176, 177, 178
BMW, 145
bot, 49
BPL (Broadband Over Powerline), 78
broadband, 27, 49, 74, 75, 76, 77, 78, 79, 80, 91, 103, 104, 155
budget, 10, 63, 108, 109, 136, 145, 150, 178, 179
Budget, vii, x, 108
budgets, 5, 10, 11, 21, 171, 172, 178, 180
business model, 1, 12, 13, 16, 22, 23, 25, 26, 27, 28, 29, 32, 34, 39, 45, 57, 58, 59, 62,
    63, 77, 82, 85, 86, 89, 95, 96, 111, 164, 173
business models, 15, 17, 29, 31, 33, 82, 167
buzz, 3, 70, 84, 144, 145, 146, 147, 148, 149, 150, 152, 159, 164, 166, 167, 169, 173

cable, 29, 75, 76, 77, 78, 79, 83
campus police, 4, 178
CAN-SPAM, 153
Cantor, Norman, 113
Carey, Michael, v
cash flow, 23, 110
CD, 24, 27, 61, 84, 85, 86, 87, 88, 89, 93, 97, 102, 141
CDMA, 82
cell phones, 9, 74, 102, 104
CEO, 3, 63, 162, 163
Charles In Charge, 64
chat, 34, 70, 103, 134, 141, 144, 147, 150, 166, 169
Chicago, 8

Clusty, viii, 50, 51
coffee, 3, 4, 21, 25, 26, 27, 28, 31, 71, 80, 150
Cohen, Stephen, 176
collateral damage, 16, 97, 110
Columbia, 8, 94
Comcast, 76, 79, 158
commercial information, 39, 164
commercial search engines, 6
commercial web sites, 38, 39, 47
commons, 82, 85, 173, 174, 175, 178, 179
communities, 7, 67, 78, 174
competition, 15, 24, 26, 27, 28, 76, 77, 79, 109, 113, 121, 148, 162
Consolidated Edison, 78
content, 1, 8, 12, 23, 24, 28, 29, 30, 31, 37, 43, 44, 46, 47, 49, 52, 53, 55, 56, 59, 65, 66, 67, 69, 72, 86, 89, 92, 102, 112, 125, 137, 138, 139, 143, 144, 153, 154, 155, 157, 158, 159, 163, 165, 170, 171, 176, 178, 179
Context Clustering Technology, 49
contextual search, 49
copyright, 7, 46, 82, 86, 92, 94, 95, 96, 97, 98, 170, 171, 172
Cornell, 8, 99, 109
critical thinking, 67, 69, 106, 165
culture, 7, 8, 9, 10, 12, 13, 16, 17, 104, 108, 161, 162, 163, 164, 166, 167, 176
Curry, Adam, 177
customer, 11, 13, 17, 18, 19, 22, 23, 24, 25, 26, 27, 28, 30, 32, 34, 35, 36, 37, 38, 42, 47, 48, 54, 60, 71, 87, 88, 100, 101, 110, 112, 113, 114, 115, 116, 117, 118, 119, 120, 121, 122, 123, 129, 138, 140, 143, 144, 146, 147, 148, 149, 151, 152, 153, 160, 162, 163, 164, 170, 171, 172, 173, 174, 178
customers, 10, 11, 12, 17, 18, 19, 22, 23, 25, 26, 28, 29, 31, 33, 35, 42, 43, 45, 57, 58, 69, 71, 75, 80, 87, 88, 89, 90, 92, 93, 100, 101, 110, 111, 112, 113, 114, 115, 116, 117, 118, 119, 120, 121, 123, 125, 126, 127, 137, 138, 139, 140, 142, 143, 146, 148, 149, 150, 151, 152, 153, 157, 158, 160, 162, 163, 164, 165, 166, 167, 168, 169, 170, 171, 172, 173, 174, 179

democracy, 15, 21, 61, 175, 179, 180
Department of Commerce, 76
Department of Homeland Security, 73
desktop search, 45, 46
DeVry Institute, 141
Dewey, John, 2, 175
dial-up, 76, 155

digital literacy, 105
digital services, 10, 118, 167, 169
distance learning, 108, 141
dorm, 3, 9, 20, 86, 88, 89, 103
DOS (MSDOS), 132, 134
downloading, 9, 82, 88, 90, 91, 95, 98, 168, 169
downloads, 27, 86, 88, 89, 91, 94, 96, 98, 103, 167, 168, 175, 177
Dunkin' Donuts, 27
DVD, 96, 102

eDonkey, 91
Educational Testing Service, 105
email, 6, 55, 62, 63, 79, 89, 91, 103, 134, 141, 144, 145, 147, 150, 151, 153, 154, 155, 156, 157, 158, 159, 160, 161, 169, 171, 177
emoticons, 63, 134
employees, 5, 12, 14, 17, 18, 19, 23, 27, 30, 71, 81, 103, 106, 110, 115, 116, 117, 118, 120, 121, 123, 124, 125, 126, 127, 132, 135, 136, 137, 138, 141, 142, 157, 164, 172, 173
entertainers, 21, 83
entertainment, 16, 25, 27, 28, 29, 71, 72, 79, 82, 83, 86, 91, 93, 94, 96, 97, 100, 102, 142, 145, 167, 173
Exxon-Mobil, 66

FAB formula, 148
Faculty, 3, 6, 7, 8, 9, 12, 13, 14, 20, 21, 22, 100, 101, 105, 108, 111, 112, 115, 116, 123, 125, 126, 166, 179
Fasttrack, 90
FCC, 65, 75, 76, 77, 78, 79, 96
Ferris Bueller's Day Off, 65
file sharing, 9, 13, 86, 91, 92, 94, 95, 97
firewall, 46
franchising, 148, 173
free, 2, 8, 24, 29, 32, 33, 34, 35, 38, 42, 43, 45, 57, 58, 63, 64, 70, 74, 80, 82, 86, 88, 90, 92, 95, 99, 104, 111, 121, 145, 146, 147, 151, 153, 157, 158, 159, 160, 162, 164, 165, 169, 174, 175, 177, 179
FTP (File Transfer Protocol), 91, 175

Garnar, Doug, v
Gates, Bill, 129
gif file, 155, 156

GLAT (Google Labs Aptitude Test), 106
Gnutella, 90
Google, viii, 32, 33, 34, 35, 36, 37, 38, 39, 40, 41, 43, 44, 45, 47, 49, 54, 57, 60, 64, 68, 98, 106, 107, 170
Google Answers, 33, 34, 35, 38
googlewhacking, 41
Gorman, Michael, 176
Graduate Record Examination, 105
graduate school, 2, 4, 101, 124
Grokster, 91, 92, 93, 95
GUI (Graphical User Interface), xi, 127, 129, 130, 131, 132, 134, 135, 139, 142, 160, 173
Gutenberg-e project, 8

Harvard, 8, 44
Hawaiian Electric Company, 78
hiring profile, 137
hits, 39, 40, 41, 47, 49, 50, 54, 55, 56, 60, 65, 86
Hoest, Bunny, v
horizon, 17, 154
hot spots, 80
html (hyper text markup language), 45, 156, 158

ICANN, 74
images, 7, 32, 37, 44, 128
indexes, 32, 46, 47, 52, 91, 92, 159
Inducing Infringement of Copyrights Act (S.2560), 96
information, 5, 6, 7, 8, 9, 10, 11, 12, 13, 14, 15, 16, 17, 18, 23, 24, 27, 28, 29, 30, 31, 32, 33, 34, 36, 37, 38, 39, 40, 42, 43, 44, 45, 46, 47, 48, 49, 50, 51, 52, 53, 54, 55, 56, 57, 58, 59, 60, 61, 62, 63, 64, 65, 66, 67, 68, 69, 72, 73, 74, 85, 89, 96, 97, 98, 99, 100, 101, 102, 103, 104, 105, 106, 111, 112, 115, 117, 121, 122, 123, 124, 125, 126, 127, 128, 132, 133, 134, 135, 137, 139, 140, 142, 143, 144, 145, 146, 148, 149, 150, 151, 152, 153, 154, 155, 156, 160, 161, 162, 163, 164, 165, 166, 169, 170, 171, 173, 174, 175, 176, 177, 178, 179
Information and Communications Technology Literacy Assessment Test, 105
information business, 13, 99, 123, 161
information content, 30, 66
information processing, 12, 105, 126, 175
information service business, 12
instant messaging, 6, 89, 103

Instant Messaging Service, 134, 145
Internet, viii, 6, 7, 8, 12, 13, 16, 20, 27, 29, 31, 32, 38, 40, 41, 42, 44, 45, 46, 48, 49, 50, 51, 54, 55, 56, 57, 58, 61, 62, 64, 67, 68, 69, 74, 75, 76, 77, 79, 80, 81, 82, 85, 88, 89, 90, 92, 93, 94, 98, 99, 100, 102, 103, 104, 105, 106, 112, 114, 135, 140, 144, 145, 148, 150, 151, 152, 155, 156, 163, 164, 165, 166, 170, 174, 176, 177, 179, 180
Internet & American Life Project, 6, 8
interviewing, 114, 135, 141, 149
intuitive, 39, 40, 45, 131, 132, 134
IP (Internet Protocol), 93, 97, 98
iPod, 71, 85, 177

job descriptions, 117, 126
job market, 101, 106, 114, 115
John Doe, ix, 86, 93, 94, 170, 175
Jones International University, 108
jpg file, 156

KaZaA, 90, 91
keystroke, 53, 61
keyword, 39, 45, 48, 49, 50, 62
keywords, 38, 40, 42, 45, 46, 47, 48, 60, 159
Klingon, 134
knowledge, 16, 37, 40, 66, 67, 68, 69, 106, 115, 126, 138, 139, 165
Knowledge Networks, 103
Kodak, 80

laptop, 20, 27, 72, 154, 166
laptops, 9, 80, 81, 166
lawsuits, 53, 78, 92, 93, 94, 97, 170
lectures, 3, 9, 20, 21, 70, 104, 178
library staff, 9, 166, 179
links, 20, 34, 43, 45, 47, 50, 52, 58, 60, 69, 89, 153, 154, 158, 160, 164
lurkers, 132, 179

Mac Paint, 131
Mac Write, 128, 129, 131
Madison, GA, 80
management, 5, 18, 19, 23, 26, 27, 30, 91, 105, 110, 115, 116, 117, 118, 121, 123, 125, 126, 127, 137, 141, 162, 163, 166, 172

marginal cost, 16, 143
marketing, 13, 30, 63, 65, 70, 82, 85, 88, 97, 100, 101, 108, 109, 111, 112, 130, 143, 144, 145, 146, 147, 148, 149, 150, 151, 152, 153, 154, 156, 157, 159, 163, 166, 167, 168, 173, 175
MCI, 77, 79
Meador, John, 5
metasearch, 41, 50
metrics, 22, 111, 169
Metro-Goldwyn-Mayer Studios Inc. v. Grokster, Ltd, 96
Mexican restaurant, 146, 147, 150
Michigan, 44, 94
Microsoft, 20, 33, 40, 41, 45, 55, 74, 86, 89, 96, 128, 139, 155, 160, 172
millennials, 9, 13, 83, 86, 102, 103, 104, 105, 107, 164, 166, 167, 177
Miller, Laurie, 5
mission, 11, 70, 162, 163, 165, 166
moblogs (mobile blogs), 176
monographs, 4, 14, 67, 69, 161, 165
Moore's Law, 104
moral and ethical issues, 82, 83
Morpheus, 95
Motion Picture Association of America, 82, 95, 175
mouse, 25, 128, 129, 131, 139, 140, 141
MP3, 85, 91, 93, 98, 177
MPAA, 82, 95, 96, 175
MS-DOS, 128
multiplier effect, 123, 124, 149
museums, 7, 13
music, 9, 12, 20, 21, 27, 32, 37, 44, 53, 55, 60, 71, 72, 81, 82, 83, 84, 85, 86, 87, 88, 89, 91, 93, 95, 96, 97, 98, 99, 102, 142, 167, 168, 169, 170, 175

Napster, 89, 90, 92, 99, 168
National Science Foundation, 51
natural language queries, 48
network marketing, 144
New York Public Library, 8, 44
news, 29, 54, 57, 61, 68, 72, 102, 113, 116, 117, 147, 159, 164
Nielsen Norman Group, 56
Nielsen Ratings, 58, 76
non-profit, 10, 17, 63, 75, 105, 108, 109
Nunberg, Geoffrey, 47

one-touch handling, 121
online games, 103
organic, 57, 60, 62
Oxford, 44

P2P, ix, x, 13, 37, 46, 82, 83, 85, 86, 88, 89, 90, 91, 92, 93, 94, 96, 97, 98, 99, 144, 145, 167, 169, 173, 174, 175, 176, 178, 179
pajamahedeen, 61, 178
password, 40, 46
PC, 16, 20, 45, 49, 72, 80, 81, 85, 90, 91, 93, 105, 126, 132, 141, 145, 154, 155, 156, 159, 160, 166, 175, 177
PDA, 72, 74, 102, 134, 154, 167
peer review, 52
peer-to-peer, 13, 46, 82, 83, 85, 94, 126, 132, 144, 149, 173, 174
periodicals, 1, 4, 9, 63, 69, 165
personal computer, 16, 93
personalized, 16, 31, 112, 113, 146, 177
Pew, 6, 8, 9, 54, 55, 56, 62, 64, 65, 69, 88, 165, 176
Philadelphia, 5, 74, 75, 77, 81
Philadelphia Free Library, 5
Pingback, 159
podcasting, 174, 177
pods, 125
pop-ups, 45
preservation, 7, 76
privacy, 45, 46, 53
private enterprise, 17, 162
private sector, 5, 10, 17, 100, 110, 111, 112, 116, 119, 162
privatization, 109
productivity, 10, 11, 13, 107, 116, 117, 118, 120, 121, 122, 123, 124, 126, 135
professors, 3, 114
profit, 12, 13, 17, 23, 30, 33, 57, 81, 89, 108, 109, 110, 111, 121, 141, 162, 164
Proxim, 81
public policy, 13, 76, 79
Purina dog food, 145
push marketing, 144, 146, 151, 152, 153

queries, 38, 41, 45, 54, 56, 65
query, 32, 38, 39, 40, 45, 46, 47, 50, 52, 54, 59, 60, 66, 68
quick printers, 23

quotation marks, 39
Qwest, 76, 77

radio, 72, 80, 85, 97, 103, 177
rank, 60
Really Simple Syndication (RSS), 159, 164, 177
RealNetworks, 96
Recording Industry Association of America, 82, 91, 175
reference desk, 6, 33, 35, 36, 37, 38, 109, 115, 163, 169
reference librarian, 38, 115, 116
reference staff, 2, 36
relevance, 59
research projects, 2, 51
RFID (Radio Frequency Identification), 122
Rhapsody, 89
RIAA, 82, 83, 85, 86, 91, 92, 93, 94, 95, 96, 97, 98, 99, 168, 170, 175
ring backs, 103
ring tones, 103
ring tunes, 103
risk, 18, 19, 20, 22, 99, 111, 130, 144, 146, 148, 156, 162
RSS, xi, 159, 160, 164, 174, 177, 178, 179
Ruckus Network, 99

sampling, 8, 56, 85, 145, 146, 158, 160
San Francisco, 74, 75
Satellite Radio, 27
satellite television, 79
Saturday Night Live, 177
SBC, 76, 77, 80
scan, 42, 43, 44, 63, 122, 140, 170
scanning, 43, 90, 122, 126, 170, 171
Scholastic Aptitude Test, 105
scraping, 160
search bar, 32, 38, 39, 41, 54
search engine, 12, 32, 33, 34, 38, 39, 40, 41, 42, 44, 45, 46, 47, 48, 49, 50, 52, 53, 55, 56, 57, 58, 59, 60, 62, 63, 64, 65, 69, 70, 86, 165
search engines, 7, 9, 12, 13, 32, 33, 35, 38, 39, 40, 41, 44, 45, 46, 47, 48, 50, 52, 53, 55, 56, 57, 58, 59, 60, 61, 62, 63, 64, 65, 67, 68, 69, 70, 90, 97, 114, 159, 164, 165, 166, 170
Search Inside the Book, 42

searchable databases, 46
searcher, 32, 38, 40, 47, 48, 50, 51, 53, 54, 60, 61
sequence, 60, 134
service, 10, 12, 13, 15, 16, 17, 18, 19, 22, 23, 24, 25, 26, 27, 28, 29, 30, 31, 33, 34, 38, 43, 44, 57, 63, 64, 74, 77, 78, 79, 80, 82, 89, 90, 99, 111, 112, 113, 114, 115, 116, 117, 118, 119, 120, 121, 123, 124, 136, 137, 138, 139, 140, 141, 142, 143, 144, 145, 146, 147, 148, 149, 150, 151, 152, 153, 157, 159, 162, 163, 164, 168, 169, 172, 173, 174
service provider, 22, 23, 143
Sharansky, Anatoly, 175
Shared Folders, 90, 93
shelf life, 72, 146
Sherman, Cary, 94, 98
Short Message Service, 103
sideways, 134, 148, 173, 174, 175
single-touch, 121
Skype, 90
SMS (Short Message Service), 103
Snap, viii, 33, 50
social skills, 142
SONY, 81, 85
Southern Company, 78
space shifting, 166
Spam, xi, 151, 152, 156, 157, 158, 159, 160
Spam filters, 156, 157, 158
sports, 29, 35, 59, 60, 72, 102, 109, 178
Stanford, 44, 94
Starbucks, vii, 25, 26, 27, 28, 174
State of Texas, 80
Stein, Ben, 64, 65
Stimulus Progression, 71
storing, 11, 177
streaming multimedia files, 20
subscription, 99, 151, 153, 159, 160, 168, 177
Supreme Court, 77, 92, 95
surveys, 59, 75, 95, 100, 102, 127, 163
syntax, 38, 39, 132
synthesizer, 84

taxpayers, 5, 108, 109
telecom, 16, 76, 78
Telecommunications Act of 1996, 75
term papers, 2, 3, 20, 64, 178
tethered, 99, 168
textbooks, 9, 67, 69, 165
The Apprentice, 141
Thumb Generation, 102
thumbs, 13, 102, 167, 169
time shifting, 166, 169
Time Warner Center, 80
TIVO, 21, 29, 96, 102, 130
Toffler, Eric, 136
Trackback, 159
training, 5, 9, 23, 37, 38, 106, 117, 123, 124, 125, 126, 127, 131, 133, 134, 135, 137, 139, 140, 142, 162, 165, 173
Trump, Donald, 141
Tucson, AZ, 81
tuition, 20, 21, 22, 38, 108, 109
tutorials, 9, 38, 70, 160, 165, 178
TV, 20, 21, 32, 58, 72, 76, 78, 79, 83, 85, 102, 103, 129, 130, 143, 147, 177

U.S. Education Department, 108
unintended consequences, 18, 20, 33, 42, 179
University of Phoenix, 108
University of Virginia, 110
URL, 32, 51, 52, 165, 179
User ID, 46, 153
utility computing, 81

variable information, 16, 32, 73, 144, 146, 153
VBMA (Viral and Buzz Marketing Association), 144
Verizon, 76, 77, 82, 93, 94, 96, 157, 158
video, 20, 21, 22, 32, 37, 44, 72, 82, 91, 93, 97, 102, 103, 124, 167, 176
vinyl, 84, 85, 86, 89
viral marketing, 144, 145, 148, 150, 159, 164, 173
viruses, 90, 152, 156
vlogs (vido blogs), 176
VOIP, 77, 81

wall paper, 103
Wall Street, 27, 57
Wal-Mart, x, 27, 72, 89, 119, 120, 121, 122, 123, 162
weather, 29, 72, 73, 135
web crawlers, 46, 47
WGBH, 177
Wi-Fi, ix, 13, 27, 73, 74, 75, 77, 78, 79, 80, 81, 104, 161, 166, 167
Wikinews, 53
Wikipedia, 51, 53
Wi-Max, 81, 104, 166
Windows, 20, 129, 139, 140
wireless, 13, 16, 73, 74, 75, 77, 79, 80, 81, 99, 104, 105, 122, 125, 166, 167, 176, 178
word-of-mouth marketing, 144
workflow, 13, 126
World Wide Web Consortium, 48

Yahoo, viii, 33, 37, 38, 39, 40, 41, 45, 47, 48, 49, 50, 55, 57, 59, 60, 63, 64, 89, 134, 157

ZDNet, 177

978-0-595-35069-8
0-595-35069-0

Printed in the United States
44146LVS00005B/24